PALGRAVE STUDIES IN THE HISTORY OF SCIENCE AND TECHNOLOGY

Palgrave Studies in the History of Science and Technology

James Rodger Fleming (Colby College) and Roger D. Launius (National Air and Space Museum), Series Editors

This series presents original, high-quality, and accessible works at the cutting edge of scholarship within the history of science and technology. Books in the series aim to disseminate new knowledge and new perspectives about the history of science and technology, enhance and extend education, foster public understanding, and enrich cultural life. Collectively, these books will break down conventional lines of demarcation by incorporating historical perspectives into issues of current and ongoing concern, offering international and global perspectives on a variety of issues, and bridging the gap between historians and practicing scientists. In this way they advance scholarly conversation within and across traditional disciplines but also to help define new areas of intellectual endeavor.

Published by Palgrave Macmillan:

Continental Defense in the Eisenhower Era: Nuclear Antiaircraft Arms and the Cold War
By Christopher J. Bright

Confronting the Climate: British Airs and the Making of Environmental Medicine
By Vladimir Janković

Globalizing Polar Science: Reconsidering the International Polar and Geophysical Years
Edited by Roger D. Launius, James Rodger Fleming, and David H. DeVorkin

Eugenics and the Nature-Nurture Debate in the Twentieth Century
By Aaron Gillette

John F. Kennedy and the Race to the Moon
By John M. Logsdon

A Vision of Modern Science: John Tyndall and the Role of the Scientist in Victorian Culture
By Ursula DeYoung

Searching for Sasquatch: Crackpots, Eggheads, and Cryptozoology
By Brian Regal

Inventing the American Astronaut
By Matthew H. Hersch

Inventing the American Astronaut

by Matthew H. Hersch

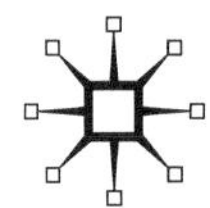

INVENTING THE AMERICAN ASTRONAUT

First published in 2012 by
PALGRAVE MACMILLAN®
in the United States—a division of St. Martin's Press LLC,
175 Fifth Avenue, New York, NY 10010.

Where this book is distributed in the UK, Europe and the rest of the world, this is by Palgrave Macmillan, a division of Macmillan Publishers Limited, registered in England, company number 785998, of Houndmills, Basingstoke, Hampshire RG21 6XS.

Palgrave Macmillan is the global academic imprint of the above companies and has companies and representatives throughout the world.

Palgrave® and Macmillan® are registered trademarks in the United States, the United Kingdom, Europe and other countries.

ISBN: 978–1–137–02528–9 (paperback)
ISBN: 978–1–137–02527–2 (hardcover)

Library of Congress Cataloging-in-Publication Data is available from the Library of Congress.

A catalogue record of the book is available from the British Library.

Design by Newgen Imaging Systems (P) Ltd., Chennai, India.

First edition: October 2012

10 9 8 7 6 5 4 3 2 1

Printed in the United States of America.

For my family.

Contents

Figures

Preface

I began collecting research material for this book in the summer of 1974; a challenging endeavor, as I was only two years old at the time. Within eight years, I had toured two field centers of the National Aeronautics and Space Administration (NASA) as well as the Smithsonian Institution's National Air and Space Museum, and had amassed a small library of primary and secondary sources on the history of spaceflight. It was not until I began a William Penn Fellowship with the University of Pennsylvania's Department of History and Sociology of Science, though, that work on this project began in earnest. My doctoral advisor, Ruth Schwartz Cowan, and committee members Robert Kohler and Walter Licht stewarded me through my research with enthusiasm and great generosity of time and spirit, and I am grateful for the many contributions of the History and Sociology of Science faculty, especially Mark Adams, Robert Aronowitz, David Barnes, Nathan Ensmenger, Henrika Kuklick, Susan Lindee, and John Tresch. My work was further aided immeasurably by conversations with members of Penn's History faculty, including Sarah Igo, Bruce Kuklick, and Walter McDougall. Susan Cerrone, Pat Johnson, and Ernestine Williams provided invaluable logistical support, and David Azzolina of Penn Libraries procured useful material for me. And, I would not have been able to pursue this project absent the assistance of the many wonderful teachers who taught me science and the historical craft.

The research and writing of the doctoral dissertation that was the start of this book project was supported by the generosity of the University of Pennsylvania's School of Arts and Sciences; I was also grateful to receive financial and logistical support from a variety of other institutions, including the National Aeronautics and Space Administration, the Smithsonian Institution, the History of Science Society, and the Society for the History of Technology. Much of the manuscript was written during the academic year 2007–08, while I held a Daniel and Florence Guggenheim Fellowship at the National Air and Space Museum. My tenure there was a remarkably productive and thoroughly enjoyable experience, to the credit of my principal advisor, Roger Launius, Space History Division chair Michael Neufeld, Will Morrison and the Office of Research Training Services, and the many other curators, museum specialists, and NASM staff members with whom I had the opportunity to work closely, including Paul Ceruzzi, Martin Collins,

James David, David DeVorkin, Jennifer Levasseur, Cathleen Lewis, Valerie Neal, Allan Needell, Hanna Szczepanowska, Thomas Lassman, Margaret Weitekamp, Frank Winter, and Amanda Young. Lindbergh fellow Robert Farquhar and Verville fellow Richard Hallion were generous with their time and knowledge, and with their recollections of the people and places I studied. Eric Long assisted me in documenting my research; Carl Bobrow, Greg Bryant, and Dittmar Geiger assisted me with access to artifacts; Jo Ann Morgan and Toni Thomas were generous with their hospitality; and Jean DeStefano, Mychalene Giampaoli, and Elizabeth Wilson and were particularly supportive of my efforts to present my work. I am also grateful to Drs. Launius, Needell, and Neufeld for facilitating introductions to several retired astronauts, who provided guidance and assistance for my research. Daniel Barry, Alan Bean, Walter Cunningham, and Jack Schmitt generously contributed their time to my study, clarifying the historical record, informing me of little-known events, and sharing their impressions and expertise.

Further archival research was conducted at the NASA Headquarters in Washington, DC, the Archives and Special Collections of Rensselaer Polytechnic Institute, and the University of Houston–Clear Lake. I am grateful to Michael Chesnes, Craig Levin, and Richard Spencer at the NASA Library, Colin Fries, John Hargenrader, Jane Odom, and Elizabeth Suckow at the NASA Headquarters Archives, and Bill Barry, Steven Dick, and Steve Garber at the NASA History Office for their encouragement of my work and the access they provided me to NASA sources. I must also thank Amy Rupert at RPI for providing access to the *George M. Low Papers*, and Regina Grant and Shelley Kelly at University of Houston–Clear Lake for providing access to the *Johnson Space Center History Collection,* as well as for their hospitality and good cheer. Web resources, including the oral history archives of NASA's Johnson Space Center, Eric Jones's *Apollo Lunar Surface Journal*, Kipp Teague's *Project Apollo Archive*, and *Great Images at NASA* also proved invaluable in the course of my work.

My research developed further while I served as the 2009–10 HSS-NASA fellow in the History of Space Science, a generous grant that provided ample support for me to revise my draft and pursue elusive sources, and for which I wish to thank Jay Malone and many others. And the book took its final form while I had the privilege to serve as Postdoctoral Fellow with the Aerospace History Project of the Huntington-USC Institute on California and the West. William Deverell and Peter Westwick provided me with the perfect opportunity to revise my manuscript during the academic year 2010–11, as well as broaden my horizons and further my study of the history of spaceflight. And I am grateful for the hard work and dedication of Chris Chappell and Sarah Whalen at Palgrave Macmillan, series editors Roger Launius and James Fleming, and several anonymous reviewers for their time, feedback, and suggestions for the improvement of the manuscript—including a new title.

Portions of this work were presented at various conferences; attendees, copresenters, and commentators including Kristin Ahlberg, John Carson,

David Haberstich, and Kendrick Oliver provided valuable feedback. Gary Hoppenstand, David Lucsko, Martin Parker, John Staudenmaier, John Waller, and the late and much-missed Ray Browne provided ample encouragement for my efforts, as well as venues for the discussion and publication of my work, and I am extremely grateful to them. Richard Barnes reviewed an early version of this work and provided many helpful suggestions. Colleagues and friends at the University of Pennsylvania and elsewhere, including (but certainly not limited to) Josh Berson, Paul Burnett, Elise Carpenter, Peter Collopy, Meggie Crnic, Brian Daniels, Deanna Day, Rachel Elder, Allegra Giovine, Mitchell Golden, Matthew Grant, Eric Hintz, Matthew Hoffarth, Andy Hogan, Andi Johnson, Chris Jones, Lindsey Plait Jones, Carmen Kroll, Whitney Laemmli, Elaine LaFay, Hui Li, Amar Majmundar, Jessica Martucci, Marissa Mika, Jonathan Milde, Mary Mitchell, Samantha Muka, Tamar Novick, Jason Oakes, Emily Pawley, Ian Petrie, Jacqueline Pravda, Katie Proctor, Bess Puvathingal, Tina Radin, Ruth Rand, Maxwell Rogoski, Joy Rhode, Corinna Schlombs, Jason Schwartz, Perrin Selcer, Brittany Shields, Nellwyn Thomas, Dominique Tobbell, Roger Turner, Jeremy Vetter, Kristoffer Whitney, Jennifer Worrell, and Damon Yarnell always encouraged me, laughed at the appropriate times, and proved the best possible sounding boards for early work. Erica Dwyer and Joanna Radin provided me with tireless support during the final months of this manuscript's preparation. Finally, production of this book would have been impossible without the kindness and support of my family (including my beloved nieces and nephews), to whom this work is dedicated.

Introduction

In 1959, a small group of American military engineer-aviators became instant heroes merely through their willingness to be hurled into space. With the Soviet Union seemingly outstripping American achievements in flight beyond Earth's atmosphere, the antidote to Communist control of the heavens appeared to lie in the courage of seven test pilots mostly unknown to the American public. Sometime soon, America's new National Aeronautics and Space Administration (NASA) announced, these brave men would be the first humans to fly into space. To support the growing needs of NASA's single-person Project Mercury spacecraft (1961–63), its larger, two-person Project Gemini vehicles (1965–66), the three-person Project Apollo spacecraft (1967–75), and the Skylab Orbital Workshop (1973–79), NASA selected, during the 1960s, new groups of astronauts roughly every other year, typically choosing a dozen or so new astronauts from a thousand or more highly qualified applicants. The 1959 and 1962 selections consisted almost entirely of active-duty military test pilots. Two civilian test pilots entered the program in 1962, and operational military pilots joined the ranks in 1963. To a press and public eager for proof that America had a future in space, the earliest astronauts fit neatly into the various roles quickly assigned to them: soldier, daring pilot, and American hero.[1]

While courageous and skilled, though, the astronauts were neither professional daredevils nor all of the nation's most respected aviators; what made them useful to NASA was what made them suspect to certain more experienced pilots: their educational pedigrees, technical training, and willingness to participate in a dangerous undertaking that would be disruptive to their military careers. As public celebrities even before they flew, the astronauts could not be easily swept aside or replaced without provoking public scrutiny, and they quickly assumed a degree of visibility that gave them power to influence various aspects of the human spaceflight program. During spaceflight's first decade, these astronauts labored to form their own distinct subculture within NASA, one with substantial authority over day-to-day engineering, training, and flight operations. They leveraged their high public profile to make early space vehicles more like the airplanes they felt safest in, and lived on the royalties of press contracts that equaled or exceeded their government salaries. Yet, like countless young officers before them, the astronauts were also "organization men": well-educated technical experts and middle-class strivers skilled in managing the often unreasonable demands of large institutions.

As celebrities in a growing government agency, the astronauts balanced a determination to meet the expectations of their superiors and the public with a desire to gain more control over their working lives. Through watchful handling by NASA managers and the American press, the actual working life of astronauts remained hidden behind a bland veneer of virtue. The often dull nature of space work, indignities of NASA's management hierarchy, fierce competition for missions, and discomforts of the civil service remained largely absent from published accounts of this new hero class. Behind their respectable personas, the private lives of some of the astronauts were as flamboyant as their working lives were tedious, yet both remained largely hidden to the public in the early years of the Space Race.

These and other facets of the working lives of the earliest astronauts remained largely hidden for nearly two decades. The tenth anniversary of the 1969 Apollo 11 lunar landing, though, coincided with a period of reexamination of the early years of America's space program. In a string of scholarly histories, memoirs, and popular works, America's astronauts emerged as imperfect human beings with an almost alien work culture. While a string of articles and memoirs cracked the door on tell-all space biography,[2] a "civilian" writer, Tom Wolfe, produced the most popular work on the all-too-human "fraternity" of America's first spacemen, *The Right Stuff*.[3] Spanning the period from early high-speed flight immediately after World War II to the end of America's first human spaceflight program, Project Mercury, Wolfe's 1979 book (and its 1983 film adaptation) chronicled the selection, training, and flight of America's "Original Seven" astronauts. In his often searing passages, Wolfe found a voice for a traditionally taciturn community: proud, unwilling to admit weakness, concerned about safety and status but not given to small talk or introspection. As noteworthy as the astronauts' courage and charisma was their arrogance, occasional coarseness, and competitive nature.

By the time *The Right Stuff* appeared on bookstore shelves, though, the culture about which Wolfe had written had already dissolved. Months after NASA accomplished Project Apollo's goal to land a man on the Moon and return him safely to Earth, budget pressures forced the cancellation of follow-on missions. Many of the astronauts who had joined the space program in the late 1950s and early 1960s to fly to the Moon (and, in some cases, stayed for the chance to fly to Mars) by the early 1970s realized that they would likely not reach either, and quit. As new groups of junior astronauts continued to fill the flight roster, NASA struggled to fill an ever-smaller number of flights with an ever-larger number of qualified astronauts. Lowest in status among them was a small group of professional scientists recruited by NASA in 1965 and 1967 against the objections of many of its veteran astronauts. Many pilot-astronauts dismissed scientist-astronauts as young, weak, and dangerously unskilled, while scientists suggested that pilots were dullards who flew more often than their skill set warranted.

Received somewhat differently were a small group of military test pilot astronauts transferred to NASA in 1969 from the United States Department of Defense's Manned Orbiting Laboratory program. Already trained as pilot-astronauts, they were immediately assigned to key support duties, but, like their scientist colleagues, held out little hope for a flight in the near term. Between July 1975 and April 1981, no Americans flew in space; by the time of space shuttle *Columbia*'s 1981 first flight all of NASA's "Original Seven" astronauts had long since retired, and few of the active duty astronauts left had actually been in space before. Most of the earliest shuttle missions were instead crewed by rookie personnel from the late-Apollo period who had accepted mundane duties, prolonged training, and the disruption of their personal lives for a chance—any chance—to fly in space. Soon, joining them aboard a new kind of vehicle, were new generations of pilot- and scientist-astronauts (both men and women), as well as an assortment of untrained private citizens to whom career astronauts reacted with a mixture of fear and condescension.

The following chapters explore the occupational experience of America's astronauts from 1959 through 1979, relating the professional culture of NASA's astronaut corps and the conflicting aspirations of its members to America's shifting ambitions in space. This story is inseparable from that of the history of the Cold War, and its changing dynamic of vigilance, paranoia, and rapprochement. Recruited at the height of US-Soviet tensions, America's astronauts, Walter McDougall describes, were merely one facet of a techno-scientific program intended to blunt the "media riot" produced by the Soviet launch of Sputnik, the world's first artificial Earth satellite, in 1957.[4] Astronauts, though only an infinitesimal percentage of those employed in America's space program, were its most visible members and the men entrusted by NASA managers to represent the program to the public. Staggeringly successful in battling the Soviet Union for supremacy of the heavens, astronauts found themselves less politically relevant once the 1960s ended.

Not quite employees and not quite "in charge," astronauts walked a fine line between autonomy and subservience, their status ensured by their unique skills and celebrity.[5] These men may be analogized to other groups of technical experts throughout history—polar explorers, nautical navigators, revolutionary architects—who parlayed their unique skills into new opportunities and temporary, disproportionate influence over historical events. The workplace boundaries that astronauts experienced were often poorly defined and in constant flux, with the astronauts called upon, at various times, to serve as skilled pilots, dutiful technicians, well-schooled engineers, and even new kinds of professional managers. As experienced pilots, astronauts were accustomed to some measure of control over their own flying machines, but as military men, they recognized the powerful authorities monitoring and directing their work.

By 1959, the era of the independent aviator had already passed, replaced with one in which the work of the pilot was mediated by new technologies of automation and control.[6] Like other workers of the twentieth century, astronauts grappled with employee-management relations, working conditions, pay, office culture, gender, and deskilling. Like other skilled technicians, astronauts risked replacement by the "dead capital" of "black boxes": "fly-by-wire" control systems, digital computers, and radio guidance systems that threatened to rob aviators of the last vestiges of personal control over their vehicles.[7] Like other twentieth-century technical workers, astronauts also found themselves subject, at various times, to different modes of control and discipline in the workplace, imposed by politicians, bureaucrats, their machines, and eventually, by each other.[8] Reflecting concerns of nineteenth-century railroad workers, junior astronauts even lamented their "capricious foremen," in the form of senior astronauts tasked with administrative duties.[9]

Committed to fly through space but unsure when they might be permitted to do so, astronauts both valued the opportunities given them and lamented the decades spent studying, training, and preparing for missions that might never leave the ground. Among the tools the astronauts leveraged in their conflicts with NASA management and each other was their status as well-schooled engineers experienced with the peculiar science of flight testing. Like other communities of American engineers during the nineteenth and twentieth centuries, astronauts struggled to build an identity around their unique technical skills. Like other engineers Robert Zussman describes in *Mechanics of the Middle Class: Work and Politics among American Engineers*, astronauts were upwardly mobile, middle-class technical workers searching for ever-greater employment opportunities commensurate with their interests, education, and growing skill sets.[10] Like Zussman's engineers, astronauts were also torn by diverse loyalties (personal affiliation, military service backgrounds) that interfered with any attempt to organize or bargain collectively with NASA's civilian leadership. And like the first generations of American engineers Monte Calvert describes in *The Mechanical Engineer in America 1830–1910*, astronauts could often be divided into two often hostile camps: highly educated scientist-astronauts and experienced "shop-floor" men, in the form of grizzled test pilots who looked upon astronauts with doctorates as sheltered academics with little practical experience.[11]

Neither valued completely for their piloting skills, nor empowered as technical workers, nor fully respected as engineers, many of America's pilot-astronauts recognized a new role for themselves in NASA: as kind of professional managers. In the same way that America's astronauts would monitor the manifold electronic systems in their spacecraft, astronaut Donald "Deke" Slayton argued, they would monitor the hundreds of NASA employees and contractors working to prepare machines and people for spaceflight. In the astronaut, Slayton suggested, NASA would find a reliable officer who could lead design teams, coordinate training activities, supervise operations, and manage NASA's considerable public relations problems. Astronauts thus provide an interesting case study in the creation of an entirely new American profession, complementing the histories of other skilled, middle-class workers who organized themselves as professionals to improve their economic and social standing.

Like the generations of American doctors and lawyers who struggled, as Magali Larson describes, to "translate ... special knowledge and skills" into "social and economic rewards," astronauts adopted the accoutrements of professionalization to distinguish themselves from other engineers and scientists working in the space program.[12] Like doctors and lawyers before them, astronauts began to define themselves less on the basis of particular hands-on skills than by a set of qualities that Andrew Abbott attributes to the modern professional, including academic training (in the form of test pilot engineering schools), abstract knowledge (the arcane rituals of flight testing), interaction with the public (in the form of press conferences and Congressional hearings), and even the creation of a set of "professional problems" to which they claimed exclusive knowledge (like lunar landing).[13] Unlike Zussman's engineers, these pilot-astronaut-managers policed the boundaries of their embryonic profession, relying upon membership standards that were, on the one hand, objective (like jet pilot training) and, on the other, so intangible that the astronauts themselves did not know what they were (though they claimed to know them when they saw them). Once selected, these men adhered to a code of service in the public interest: like other professionals, one veteran instructed new apprentices; an astronaut was an astronaut "24 hours a day," with a fiduciary duty to space exploration that transcended all other obligations. Examination of the work of space professionals also sheds light upon on the complex art and science of personnel management. Throughout the 1960s in particular, a handful of senior men possessed almost total authority to crew flights and assign junior pilots to support work, a responsibility fraught with controversy and frequent acrimony among astronauts.

The experience of this unusual group of professionals speaks not only to labor history, but also to the history of technology and "big science" during a tumultuous period in American history. Neither completely socially constructed nor wholly predetermined, the technology of spaceflight evolved during the 1970s to meet changing needs and constraints. Like other large national endeavors of the Cold War, human spaceflight persisted into the 1970s even after most public enthusiasm for it had subsided. The 1970s saw the advent of a variety of new technologies in computing, materials, and medical science, but for NASA, the decade was one of fiscal austerity and bricolage, as the agency struggled to exploit existing hardware in new ways to accomplish diverse exploration goals. NASA's investment in Apollo-era launch vehicles and the scarcity of funds available for innovation at first created a path dependence that tied NASA operations to equipment already in existence, or which could be readily modified from it. Astronauts found themselves called more and more to perform actual engineering work in space, improvising as they went along. Decisions made during the late 1970s about the design and use of the space shuttle in particular dominated American astronautics for more than three decades, as the nation struggled to recoup investments in machinery and manpower.

Though NASA's human spaceflight program struggled principally with engineering problems, to a larger public, human spaceflight, like previous government-sponsored expeditions, was a scientific endeavor critical to national security. By sending men to the Moon in the way Lewis and Clark had explored the Louisiana Purchase, America would protect its skies by venturing into them. Along the way, it would demonstrate its intellectual might by gathering important knowledge about extraterrestrial worlds for the benefit of people everywhere. In many ways, the young scientists and technologists who flew America's first space vehicles shared much in common with earlier generations of Americans who exploited educational, military, and institutional opportunities to prosper at the forefront of exploration.[14] Like other grand engineering projects, spaceflight exhibited strong ties to military institutions and personnel, and appealed to the desire of an energetic nation to expand its boundaries into unknown frontiers. Yet, NASA's pilot-astronauts (and even the public) approached science warily, and scientist-astronauts with suspicion, even hostility. And while born as a "big science" crash program (comparable to America's effort to develop nuclear weapons), human spaceflight continued into the 1970s as a new kind of technological infrastructure too large to abandon but with an unclear mandate for the future. NASA's astronauts, meanwhile, found themselves becoming like the thousands of other government engineers employed in maintaining the edifice of modernity: a strategic technical reserve maintained to preserve American preeminence for an unknowable future.[15]

Lastly, this work seeks to reconcile the cloistered labor experience of astronauts with the celebrity that accompanied them, which they and NASA managers struggled to control and exploit. During the postwar period, almost no other engineering community in the United States found its actions subject to such scrutiny, or was so dependent upon the vicissitudes of American public opinion, represented through media outlets, voting booths, and the ever-watchful eye of Congress. NASA's efforts to control its astronauts paled in comparison to its effort to influence fickle popular sentiment about spaceflight. Astronauts, keenly aware of how they would be perceived, and how they might use those perceptions to their advantage, used their leverage to assert some measure of control over their work. Rather than slavishly cowing to institutional or individual image makers, though, popular culture created its own, alternate vision of spaceflight, criticizing and punishing NASA when it diverged from public fantasies.

As Asif Siddiqi has written, Soviet spaceflight research of the 1950s was greatly shaped by Russian philosophy and popular fascination with the cosmos.[16] In America, novels, movies, and television about astronauts played a similar role. The work product of the astronauts of the 1960s—narratives, Moon rocks, and, as Sheila Jasanoff and others have written, photographic imagery—inspired popular culture that, in turn, shaped space policy.[17] In the early 1970s, a more media-savvy NASA sought to align its efforts with the ideals of environmentalism, equal opportunity employment, internationalism, and utopianism—ideals increasingly expressed by NASA's astronauts and the science fiction they inspired. In this new cultural landscape, the boundary between the true and the imaginary became porous, as people inside and outside NASA looked to science fiction to chart America's course in space. Ultimately, a network of actors—technical imperatives, political authorities, institutions, individuals, patterns of labor, and popular culture—combined uneasily, and in constant flux, to shape American space policy.

Chapter 1 describes the occupational culture established by the "Original Seven" astronauts of Project Mercury, the world only hinted at in *The Right Stuff.* Junior officers culled from the ranks of America's military test pilots in 1959, astronauts emerged from an insular, almost monastic engineering culture inured to extreme risk. Indifferent to danger, seemingly lacking in self-awareness, and obsessively ambitious, astronaut candidates evinced attitudes that simultaneously "impressed" and frightened NASA psychiatrists. Suddenly thrust into the public spotlight, though, the Original Seven combined shrewd self-promotion with often reckless private behavior. Despite occasional professional and personal stumbles, their culture remained largely unchallenged for half a decade, as successive groups of pilot-astronauts accepted and reinforced the standards of conduct, professional outlook, and workplace dynamics of the Original Seven.

Emboldened by the success of Project Mercury, its ranks expanded by President Kennedy's Moon directive, the NASA astronaut corps enjoyed, in the mid-1960s, a period of optimism manifested in the successes of Project Gemini and the accelerating momentum of the Apollo lunar program. For astronauts, working at NASA in this period involved equal parts flying, engineering, and public relations. Chapter 2 examines the activities—in space and on Earth—that filled the astronauts' working lives at the heyday of their prominence, and the growing dissonance between public perception of their work and its occasionally grim realities. On the ground, astronauts commuted from their homes in Houston, Texas, by supersonic fighter jet to the launch sites, classrooms, factories, and public events where they designed for, learned about, trained for, and promoted human spaceflight. While senior astronauts struggled to juggle scarce personnel resources and maintain NASA's breakneck schedule, junior astronauts jockeyed for flying opportunities. In space, astronauts tried desperately to satisfy their own personal ambitions, control their wonder and fear, manage the expectations of their colleagues, and avoid career-ending mistakes.

Chapter 3, spanning the years 1965 to 1972, explores the first major threats to the workplace culture of America's astronauts, in the form of professional scientists recruited by NASA to enhance the public profile of its scientific efforts. Many of the scientists selected by NASA in 1965 and 1967 proved a poor fit within the existing cultural norms of the pilot-astronauts, who described the new arrivals as slightly less useful in the cockpit than NASA's "secretaries." Echoing Helen Rozwadowski's work on maritime scientists of the nineteenth century, NASA's veteran pilot-astronauts regarded scientists as weak, useless, and unlucky members of any space crew.[18] Behind the teasing, though, lay concerns that the arrival of nonpilots at NASA endangered mission success and threatened the hard-won influence and reputation of America's space heroes.

Chapter 4 examines the occupational life of astronauts in the years between the last Apollo Moon landing in 1972 and the rollout of the first space shuttle, *Enterprise*, in 1977, emphasizing the often bitter departure of veteran astronauts and the struggle of their younger, lesser-known colleagues to find seats in the last American spaceflights of the 1970s. In contrast to the scientists, NASA's last group of pilot-astronauts selected during the 1960s, a small group of respected air force pilots and navy aviators transferred to NASA in 1969 from the military Manned Orbiting Laboratory program, found few flying opportunities but experienced a professional culture that respected them as well-trained journeymen. This chapter focuses on the experience of these groups of spacemen, including the experiences of the 1973 Skylab crews, for whom rides in space became tedious and unglamorous, but just as dangerous as ever. Having enjoyed unprecedented celebrity and authority in the 1960s but now no longer in the spotlight, astronauts of the 1970s adjusted to a work culture that placed them more firmly under the control of NASA management and demanded from them new skills of negotiation and adjustment.[19] Like earlier workers who endured a shift from the freewheeling market economy of late nineteenth century to an economic landscape dominated by large corporations in the early-twentieth, astronauts in the 1970s adjusted to a new NASA in which their goals were aligned more closely with those of the agency and their independence was sharply curtailed.[20]

In the early 1970s, a deeply conflicted spaceflight community struggled to squeeze more and more from a space program able to afford less and less. In order to maintain public funding of spaceflight through the 1970s, NASA would need to find new rationales for exploration and galvanize new sources of public support. Chapter 5 describes how a new, more media-savvy NASA emerged in this period, eager to capitalize on popular enthusiasm for science fiction and more able to exploit astronauts as spokesmen for its efforts. With the space shuttle, America's replacement for the Apollo spacecraft, NASA would seek to satisfy diverse public constituencies for spaceflight. Yet for astronauts, the shuttle's amorphous purpose, questionable management, and uncertain crewing proved a source of increasing concern.

Chapter 1

"Project Astronaut"

Pondering names for America's first human spaceflight program in the fall of 1958, Robert Gilruth, Chairman of the Space Task Group (STG) of the young National Aeronautics and Space Administration (NASA), suggested "Project Astronaut," but others at NASA feared that this name would draw undue attention to the personalities of the hand-picked aviators whom NASA's army of engineers would soon blast into space.[1] Instead, Abe Silverstein, NASA's Director of Space Flight Programs, suggested "Project Mercury," an evocation of the wing-footed Roman god and a continuation of the American custom of naming rockets after characters in Greco-Roman mythology. From the earliest days of America's human space program, the role of the pilot was the subject of substantial controversy. Intended to serve as representatives of a large national technical endeavor, the astronauts—and not NASA's scientists, engineers, or managers—quickly became its public face, and popular heroes. The earliest American astronauts assumed greater prominence than many in NASA had expected, and structured their working lives not only to enjoy the benefits of their celebrity, but also to protect themselves physically and professionally in an endeavor that they hoped would mark the beginning of their careers, not the end of them.

America entered the Space Age in the fall of 1957 with its military services and government laboratories already studying the requirements of piloted flights into Earth's orbit, and already employing the professionals who would crew its new space vehicles. Long challenged with the role of proving out dangerous new flying machines, military test pilots soon emerged as the most convenient pool of potential astronauts, and accepted participation in the burgeoning space program despite its apparent indignities. Though often drawn to the allure and danger of flying, these men were far more studious and responsible than they seemed—professional engineers comfortable with working in large organizations. The early test pilots were able, ambitious, and aggressive men, but they were neither entrepreneurs nor heroic inventors, nor captains of industry. Rather, they were elite technologists who, despite certain unique personality traits and uncommon abilities, had succeeded in large organizations known for their limited autonomy and highly circumscribed standards of conduct and performance.

Upon their selection, astronauts balanced a need to fulfill the expectations of their civilian managers with a desire to increase personal control over various aspects of their work, especially those pertaining to safety. Military test pilots brought experience and expertise to the development of space vehicles and were well-suited to the task of flying them. Throughout the early years of the space program, though, successive classes of these pilot-astronauts encouraged the development of quasi-military institutions within NASA and cultivated spacecraft so sophisticated that only they could fly them. By the end of NASA's first five years, the seven men who comprised the nation's astronaut corps had established an internally managed suborganization within NASA with substantial, if carefully circumscribed, control over many of its own affairs. This small group of astronauts not only managed to recast their relationships with NASA's engineers and physicians, but also established the norms and traditions that would characterize the astronaut office for the next 15 years.

"As the Service Requires"

Of all of the individuals proposed to fly into space, military pilots had, by 1957, emerged as the most likely candidates. In the mid-1950s, America's airplane test pilots appeared to be edging toward space above the dry lake beds of Edwards Air Force Base, California where, for nearly 15 years, aviators flying under the auspices of the US army, air force, navy, and civilian National Advisory Committee on Aeronautics (NACA) had "wrung out" a series of experimental aircraft intended to push the known limits of speed and altitude. In 1947, US army air force's Major Chuck Yeager, a decorated World War II ace, piloted Bell Aircraft's stubby, rocket-powered X-1 faster than the sound waves that the vehicle itself had produced in flight, a "magic" speed that produced shock waves so violent they had shattered previous airplanes into pieces almost too small to be found. Dropped from a carrier aircraft over the desert, Yeager rocketed the X-1 to 43,000 feet and demonstrated that the infamous sound "barrier" was penetrable with proper aircraft design and handling. In the years that followed, successive jet- and rocket-powered craft doubled and tripled this speed, at altitudes so lofty that the flaps and stabilizers of conventional flight controls could no longer grab enough air to steer the craft. By the late 1950s, balloons had carried humans to altitudes in excess of 19 miles, and the latest rocket-powered craft, North American Aviation's X-15, was poised to send a single pilot on brief arcing trajectories to altitudes in excess of 50 miles and at speeds greater than six times the speed of sound.[2] At such heights, virtually all of the Earth atmosphere would lie below the pilot; steering the rocket craft with small thrusters, the pilot would enjoy a few tantalizing moments of spaceflight before gliding the craft to a safe landing.

X-plane work, like most flight-test programs for civilian and military aircraft, was a research discipline requiring formal training equivalent to several years of undergraduate and graduate engineering education.[3] Flight testing emphasized not radical handling and death-defying stunts, but a slow, incremental process through which the characteristics of new aircraft were evaluated in a series of choreographed maneuvers, followed, as former test pilot and NASA astronaut Mike Collins later recounted, by laborious data analysis:

> The general pattern was: first, by classroom lectures, to learn the theoretical aspects of a particular type of test, then to make one or more flights trying out the new technique, and finally to analyze the reams of test data acquired during the flights. This last part, the reduction of thousands of bits of information ("data points") into an intelligible report, complete with charts and graphs, was a tiresome, time-consuming process. ... After each flight, the developed film and oscillograph paper were delivered to us; and weekends and nights would find us hunched over a desk calculator or peering at a film projector, trying to reduce this overwhelming amount of information into a terse report.[4]

To outsiders, the test pilots seemed to be foolhardy men who lived life "on the edge." To the pilots, though, their work was as dull as it was dangerous, and required a more diverse set of skills than those that had equipped previous generations of combat pilots and barnstormers.

The test pilots who would become America's first astronauts arrived in this community just as new technologies were replacing hands-on flying with greater integration of electronic systems into flying vehicles. "In earlier flight," one 1959 air force publication concluded, flying entailed a "high level of psychomotor ability...for example, in the coordinated manipulation of stick, rudder bar, and throttle. At present, advances in mechanical systems have reduced these functions essentially to those of information processing and decision making."[5] Confronted with new forms of propulsion and automation threatening human piloting, a group of leading test pilots, in 1955, organized a professional society—the Society for Experimental Test Pilots (SETP)—to address the place of man within increasingly complex mechanical systems.[6] These discussions, though, did not take the form of labor unrest: indeed, minutes of the group's first meeting included the consensus determination that SETP's "primary purpose" would be the development of safety and survival equipment. Apparently concerned that the group would be mistaken for a trade union, the founders also determined to "obtain industry sanction" for their meetings, resolving to prepare an explanatory "brochure" about the organization for "the various aeronautical establishments" employing them.[7] Test pilots, throughout their history, avoided efforts to bargain collectively with their employers, though they would occasionally reassert their continued importance in flight test engineering.

The most obvious threat to the tests pilots' skill-set was the unpiloted rocket, lofted into space using radio links and computerized guidance.

By January 1957, it was obvious that rockets would soon accelerate instrument packages (and, eventually, human beings) into space not for mere moments, but for hours, as satellites in orbit around Earth.[8] To achieve this feat, the "spacecraft" would need to achieve not only height, but speed: the nearly 18,000 miles per hour at which an object's trajectory would follow the curvature of the Earth, causing it to "fall" around the Earth without the input of any additional propulsion. The enabling technology to achieve such speeds was not the jet engine or the X-plane, but the large, liquid-fuel military rocket, which the United States and Soviet Union rushed into developing in the 1950s as an intercontinental delivery vehicle for nuclear weapons. By the time the intercontinental ballistic missile (ICBM) appeared in the Soviet Union and United States in 1957 and 1958, such vehicles possessed the thrust to either lob a warhead across an ocean or accelerate a payload into Earth orbit. Used as a booster, such a rocket could even lift a piloted vehicle into space.

Many within the NACA and the air force hoped that the X-20 "Dyna-Soar," a successor to the X-15 rocket plane, would accomplish this feat by the 1960s. Soaring into orbit atop a modified Titan ICBM, the X-20 would maneuver under human control before decelerating, descending, and gliding to a landing.[9] The first American piloted spacecraft program to reach hardware stage, the X-20 never flew; months after the program's inception, *Sputnik 1* sailed into Earth orbit atop a missile that, with the addition of an upper stage, might carry a small, piloted capsule, returning safely to Earth using breaking rockets and a parachute at the conclusion of the flight. Even though NACA engineers preferred to continue reaching toward space in rocket planes, by 1957, the NACA's leadership recognized that the urgency created by *Sputnik* required the United States to abandon the relatively heavy and complex rocket plane technology for lighter piloted capsules more easily adapted to existing ICBMs.[10]

The launch of *Sputnik*, though, had not taken American spaceflight proponents by surprise. Formal discussions in the United States about the medical qualifications for future spacefarers had begun as early as 1952.[11] In March 1955, the air force convened a panel of test pilots in Washington, DC to establish medical selection criteria for future space pilots. The March 1955 panel, chaired by Air Force Office of Scientific Research head Brigadier General Don Flickinger, produced no strict physical standards, declaring instead that future craft would likely require traits common to existing test pilots. The panel also concluded that pilot motivation was the factor most critical to flying success, and that educational credentials were largely irrelevant.[12] Later, Flickinger summarized the traits required of America's first "satellite pilots," who he insisted would be, by necessity, a "thoroughly picked group," "calm and confident," and so modest as to seek no glory in the endeavor:

> We must screen our personnel to find those who are experienced observers, who have demonstrated qualities of coolheadedness and resourcefulness in

> tight situations, who have good intelligence, who can tolerate various extremes of physical punishment, and who are stable, calm, and confident. We must reject those who, although able to give a good account of themselves, do so primarily to prove something to themselves or the world. Then we must narrow this select pool of potential satellite pilots down to those six to ten who are best fit to undertake such a mission.[13]

Further air force studies refined the requirements, attempting to elucidate the selection criteria for space pilots and the potential physical stresses of orbital flight.[14] Some of these standards later found their way into calls for astronaut applicants (like a requirement of 1000 hours of jet piloting experience), while others spoke to persistent biomedical concerns that seemingly innocuous aspects of the space environment might prove dangerous to human functioning. Among these were the "withdrawal of smoking, cokes, coffee, and snacks"—a concern, as virtually all of the Project Mercury astronauts smoked. Some physicians also feared that the sudden "separation" of "familiar" social "supports" in space would lead to feelings of "exhilaration" and "omnipotence" in some pilots, and depression in others. More predictive of future astronaut selections were conjectures that candidates would likely endure comprehensive medical and psychological examinations, and that once selected, they would likely demonstrate "ego involvement" in the flight test program and "be identified with the project virtually from the blueprint stage."[15]

In June 1958, the air force made an informal preliminary selection of potential crew members for its Man-In-Space-Soonest (MISS) project, which would attempt to orbit a piloted capsule launched atop a ballistic missile. The air force appears to have employed no formal criteria in the selection, choosing the names on the basis of reputation alone and without consideration of biometric or other data. This list comprised nine respected USAF and NACA test pilots, including a young civilian NACA pilot with a naval background, Neil Armstrong. When the newly chartered NASA absorbed the old NACA and assumed control of "manned" (later, "human") spaceflight in August 1958, the air force cancelled the MISS project, but air force research remained influential in the selection of astronauts for NASA's early years, as the agency struggled to establish its own facilities and personnel and organize a new civilian human spaceflight program.

For NASA's early astronauts, Robert Rowe Gilruth, a Minnesota-born aeronautical engineer who had joined the NACA in 1937, became the most important face of NASA management and the individual with the most direct control of the astronauts' working lives. After World War II, Gilruth had risen rapidly in the NACA's ranks, becoming one of its highest-paid managers by 1948. Shortly after NASA's formation in 1958, administrator T. Keith Glennan assigned Gilruth to chair the STG, NASA's fledgling human spaceflight initiative based in a nondescript office building near the NACA's Langley Research Center in Virginia. If German émigré rocket designer Wernher von Braun epitomized America's rocket-making leadership before and after NASA's formation, Gilruth, with a much lower public profile, spoke for its test pilots, and managed the problems of human flight, including the creation of spacecraft and recruitment of the people who would fly them. The three dozen engineers Gilruth led defined the parameters of Project Mercury and managed its development.[16]

It was not certain in late 1958, however, that pilots would fly first. Without a clear idea of what its astronauts would do in space or how well they could perform, the STG was initially vague on its requirements for NASA's new astronaut corps.[17] Early space vehicles resembled high-altitude balloon gondolas and underwater bathyscaphes more than airplanes, and a wide variety of personnel—divers, balloonists, submariners, mountain climbers, scientists, physicians—seemed to possess the requisite fortitude to operate them.[18] Early in the 1983 film version of *The Right Stuff*, Tom Wolfe's novelistic account of the early years of the Space Race, two hapless NASA bureaucrats (played by Jeff Goldblum and Harry Shearer) attempt to pitch potential astronauts to a skeptical President Dwight Eisenhower. Mocking debates within NASA at the time, the proposed candidates include a variety of circus performers and professional daredevils whose principal qualifications appear to be their comfort with heights and the fact that they "already own their own helmets."[19]

In November 1958, the STG requested permission from the Civil Service Commission to hire 40 "'scientific specialists' to participate in 'special research activities. . . .'"[20] The recruitment announcement, dated December 22, sought "Research-Astronaut Candidates," more broadly conceived than those that the air force had proposed, to be hired and paid as civil servants at the GS-12 to GS-15 level. While the selection limited applicants to males aged 25 to 40, in good health, with college degrees and several years of postgraduate scientific or technical work, these individuals did not need not be pilots, or even military officers. Rather, NASA defined qualifying professional experience as command experience in balloons or submarines, PhD-level research in science or engineering, medical or psychological training, and equivalent forms of technical expertise. NASA's final requirement was that candidates demonstrate experience in "hazardous, rigorous, and stressful experience," including, but not limited to, test piloting, "mountain climbing," "wartime combat" or other situations "whether as occupation or sport." No one form of intrepid explorer would fare better than any other in the selection, the announcement suggested, or would be more qualified to fly in space.[21]

Following publication in the *Federal Register* on December 9, 1958, though, NASA quickly pulled the announcement as STG members and aviation physicians began leaning toward military test pilots, who NASA could recruit and select outside of normal civil service procedures.[22] Robert Gilruth, in particular, was enthusiastic about this idea. Military aviators, Gilruth noted, were already technically knowledgeable, familiar with the stresses of speed, altitude, and risk, and were accustomed to discipline. They also, literally, possessed their own helmets, in the form of familiarity with pressure suits and other high-performance flight hardware sure to be required in the space program. While many civilian test pilots likely possessed the requisite skills as well, military pilots could be contacted and interviewed more quickly than any other group; they also had security clearances and were, presumably, willing to move to unpleasant, far-flung locations with few questions asked. NASA might not even need to pay them; military pilots were already on the government payroll and the pilots' service branches could detail them to NASA for as long as NASA required their services. (The detailing of military pilots to NASA later provided NASA management with a convenient mechanism by which to discharge astronauts who proved unsatisfactory.)

To the surprise of Robert Gilruth, Keith Glennan, and deputy administrator Hugh Dryden, President Eisenhower was enthusiastic about the idea of military test pilots serving as astronauts in the new civilian spaceflight program, approving the policy orally in a matter of minutes during an informal conversation at the end of December 1958.[23] "They are in the service to do as the service requires of them at various times," Eisenhower insisted to Gilruth, suggesting both that the pilots had earned the government's indulgence, and that they were, perhaps, obligated to assist the nation in this critical undertaking. Indeed, Eisenhower continued, the test pilots' record of service demanded that they be given "a chance to volunteer if they wish."[24]

NASA established its second set of astronaut selection criteria in a meeting at NASA Headquarters in Washington, DC on January 5, 1959, which was attended by several NASA and military insiders, including Gilruth, Flickinger and Dr. W. Randolph Lovelace, II, chairman of the NASA Special Advisory Committee on Life Sciences. After leading the air force's aeromedical laboratory at Wright Field during World War II, Lovelace had retired to become a government consultant, establishing a private clinic near Albuquerque, New Mexico, that eventually served as an examination site for astronaut hopefuls.[25] Instead of circus people or conscripts, America drew its first astronauts from the graduates of the test pilot schools at Edwards Air Force Base in California and the Naval Air Station at Patuxent River, Maryland.

Military services would nominate candidates directly to NASA; as service regulations prevented army pilots from flying fixed-wing aircraft, though, the few army names offered failed to meet even the most basic standards for consideration. Flickinger and Silverstein attempted to resolve this problem by approaching the army for names of additional candidates, none of whom ultimately proved qualified.[26] Despite the army's inability to participate, NASA's decision seemed to guarantee an all-military astronaut corps for the foreseeable future, given the large number of available pilots from other military service branches and the apparently "finite" duration of Project Mercury.[27] Robert Gilruth and future Apollo program director George Low solicited civilian NASA pilots at various field centers for their interest in Mercury, but none volunteered, "although several expressed interest in joining the Project at a later date." These were the only civilians NASA considered, although Low recounted in an April 1959 memorandum of having received, despite the secrecy of the selection, "approximately ten letters" from civilians requesting consideration as astronauts, some "sincere" and "several…obviously from cranks."[28]

That none of the astronaut candidates would be female, non-Caucasian, or foreign-born was so obvious to the Selection Committee as to go unstated. Once NASA had decided on military aviators, the ethnic and gender characteristics of the astronaut corps were guaranteed: women, African Americans and ethnic minorities were entirely absent from the graduates of the nation's two test pilot schools in 1959, and even many talented White aviators lacked the 1500 hours in command of jet aircraft required of all astronaut applicants. The selection process ensured not only the early racial and gender homogeneity of the astronaut corps, but its emphasis on formal training and institutionalized "overcredentialism."[29] The initial stringency of the astronaut selection criteria was a source of pride for NASA during its early years, when it had little to show for its efforts.

Though later critics likened America's first astronauts to "lab rats," NASA, under Eisenhower's influence, sought promising career aviators with substantial time in the cockpit, not the nation's youngest or most docile pilots. As one examiner later noted, astronauts would be responsible for more than merely providing biomedical data, operating experiments, or even flying. NASA hoped its astronaut-candidates would demonstrate "evidence of sufficient drive and creativity to insure positive contributions to the development of the vehicle," and therefore sought out promising, well-qualified flyers with substantial engineering training and a willingness to accept temporary reassignment from promising careers to participate in a new, untested organization.[30] The unreliability of rockets of the era suggested a particular need for men comfortable with the extreme hazards involved in rocket flight. In one oft-cited statistic, test pilot work killed more than one-fourth of those undertaking it; for such men, death in the line of duty, however dreaded, was unremarkable.[31]

Unable to maintain the secrecy of its selections, NASA would eventually confront public and private questioning about why America's civilian space program was unable to find any nonmilitary candidates suited to space work, and how it could effectively manage men whose loyalties would appear to lie with their respective service branches.[32] NASA would, Silverstein wrote to Glennan on January 29, 1959, face the challenge of "making these men civilians": both publicly palatable and amenable to NASA discipline. To resolve this problem, Silverstein suggested that the pilots selected be detailed to NASA on "3-year detached duty status," that they never wear their uniforms, and that they be "responsible to NASA, and not to the military services," and that NASA pay their salaries. "This proposed procedure would," Silverstein wrote, "make these men NASA employees, even though they are still in the military."[33] Ultimately, Silverstein's procedures would do far more: create a collective of pilots not fully beholden to either NASA or to their military branches, and able to define their own workplace culture.

Having arrived at an all-military astronaut corps, NASA also faced the problem of convincing military men to transfer to a civilian program. "I believe that the question of status is of the utmost concern to a military man," Silverstein wrote, "the pilot selection program may well be compromised unless we can tell the first group of men what their eventual status will be."[34] Indeed, to many of America's would-be astronauts of 1959, Project Mercury at first appeared to be only the latest in a series of high-speed, high-altitude flight test programs, distinguished primarily by its civilian management and the high degree of unreliability associated with its hardware.

Pressured to meet cost and schedule objectives, Project Mercury (like the air force's abortive MISS Project) would utilize an Atlas ICBM (the only launch vehicle in the United States arsenal powerful enough), to hurl a 3000-pound remote-controlled capsule into orbit; the capsule's "pilot" would do little more than monitor instrumentation and survive, demonstrating the biomedical feasibility of human spaceflight.[35] The Atlas itself had been designed around the high failure rate associated with liquid-fuel rocket motors of the era, with three large engines that fired, not in sequence, as in "staged" rockets, but simultaneously, so that ground controllers could verify that they actually worked before liftoff. During its orbital coast, the Mercury capsule would require little maneuvering; during its return to the ground, aerodynamic and ballistic forces acting on the capsule would ensure its proper orientation, slow it down, and effect its descent.[36] Preliminary Mercury-Redstone flights would be powered by a smaller single-stage missile, the capsule would not even achieve orbit, following instead a suborbital ballistic trajectory little different than that of a cannonball.

The earliest air force proposals to use ballistic missiles to propel piloted vehicles into space had led to charges that such an endeavor would be little more than a circus stunt.[37] Privately, the response of some of the nation's most accomplished test pilots to Project Mercury's call for rocket passengers was one of disdain. Rocket planes like the X-15 (which achieved its first powered flight in September 1959) were human-controlled from takeoff to landing and, unlike Mercury capsules, enjoyed smooth, controlled transitions from atmospheric to exoatmospheric flight. Anticipating flights in the X-20, test pilots Armstrong and Milt Thompson simulated human control of the launch vehicle from liftoff through orbit, proving that humans could fly it into Earth orbit by their own hand.[38] "We believed we would fly into space," Thomson later wrote. "The pilot would have control of the vehicle during launch, and when the mission was over, he would land on the Edwards lake bed, just as we did in the X aircraft."[39] During launch of Atlas, though, a bank of computers on the ground would steer the missile and insert the Mercury capsule into Earth orbit. The capsule would not even "land with dignity" but splashdown in the ocean, dangling from a parachute.[40] To Yeager, the laboratory primates who would occupy the first Mercury capsules were the real test pilots, and the astronauts who followed them merely better-qualified replacements, forced to squeeze into an automated cockpit formerly occupied by test chimps and littered with monkey droppings.[41]

To many of the pilots who competed successfully for Project Mercury, the program initially appeared to be a poor career move, especially since it would be run, unlike the X-plane program, by civilians.[42] Senior pilots were particularly skeptical; in some cases, military superiors balked at the invitations extended to their subordinates: air force pilot L. Gordon Cooper Jr.'s commanding officer at Edwards Air Force Base warned Cooper to avoid the "idiotic" project, but Cooper did not take the advice.[43] Naval aviators Walter "Wally" Schirra Jr. and Alan Shepard Jr. feared that Mercury would offer few piloting opportunities and would jeopardize their promising naval careers.[44] Schirra hoped for command of a ship, and wondered whether his participation in Project Mercury would undermine his chances of receiving it.[45]

Other flyers felt that better spaceflight opportunities existed elsewhere. Armstrong, was, by 1959, already committed to the X-15 project and saw little benefit in abandoning a promising flight program for one some regarded as a poorly conceived stunt (Figure 1.1). "I was flying the X-15 and I had the understanding or belief that if I continued, I would be the chief pilot of that project," Armstrong later recalled. "I was also working on the Dyna-Soar, and that was still a paper airplane, but was a possibility.... It wasn't clear to me which of those paths [would be best]."[46] Space-minded air force pilot Edwin "Buzz" Aldrin, angling for follow-on spaceflight programs, declined test pilot training for postgraduate study in astrodynamics at MIT, completing a doctorate in time to join the NASA astronaut corps in 1963. Air force test pilot Donald K. "Deke" Slayton, though entering the astronaut selection process, lamented that he would not reach space in the X-20, a project

Figure 1.1 NASA pilot Neil Armstrong stands next to X-15 ship #1 after a research flight (NASA photo)

that seemed to him a more natural evolution from aviation technology than Project Mercury's precarious rocket experiment.[47]

"Strong Intimations of Psychopathology"

Selection of NASA's first seven astronauts began with a review of the records of 508 test pilots from the navy, air force, and marines, by a committee comprised of two NASA managers (one of them a test pilot), a flight surgeon, and two psychologists. From these records, the Selection Committee chose and ranked 110 pilots to invite to Washington for a series of interviews in February 1959. Although charged with selecting men for a short term research program, NASA's Selection Committee possessed an additional agenda: to gain data about applicants that would be useful in future selections. (Of particular interest was the question of whether America's junior test pilots made better astronauts than its more senior ones.) [48] All 69 of the pilots who eventually received the invitations attended briefings—so many, in fact, that NASA decided not to inform the remaining pilots on their selection list.

Whether the pilots regarded these "invitations" as voluntary, though, is unclear. Among those who accepted NASA's offer to enter the selection process, many later claimed to have done so involuntarily.[49] Wally Schirra wrote that he had never actually "volunteered" to participate in Project Mercury; rather, he, like dozens of other officers, was ordered to attend the ostensibly voluntary briefing in Washington DC—the first stage of the astronaut selection process.[50] Naval aviator Malcolm Scott Carpenter later recounted that it was his wife who had accepted NASA's offer to interview, as Carpenter was at sea at the time. The early astronauts, though, often denied interest in a program for which they, in fact, competed furiously. Doing so both excused their participation in a professionally questionable endeavor, and, later, provided the kind of amusing anecdotes that sold the program to journalists eager for human-interest details.[51]

Following the 69 interviews, only 16 pilots declined to continue (NASA preliminarily rejected another 21). On the basis of the interviews, George Low concluded that the remaining applicants showed such a "high rate of interest" that he expected "few, if any, of the men will drop out during the training process."[52] Whatever their initial reservations, candidates aggressively sold themselves, convinced, as Schirra later noted, that if anyone was going to be launched into space, it might as well be him.[53] Had any of these men later demonstrated, during the subsequent weeks, any lack of commitment to the project, NASA would have likely dropped them. One candidate, in fact, was ranked just below the men ultimately selected merely for having confessed to examiners that he was "not entirely sure that he desired to continue on in Project Mercury."[54]

For younger pilots, Project Mercury presented an opportunity to avoid some of the more demoralizing sacrifices of military life and leap ahead of colleagues in prestige, rank, and responsibility. Well-educated, ambitious, but not well-paid, military aviators were college-trained professionals who endured austere military housing, frequent relocation, and a fair amount of financial insecurity in the hope of securing the most desirable flying and command assignments. Among aviators, advancement opportunities had traditionally accrued to those flying the most high-performance aircraft, as did status and prestige within the service. Such flying was also viscerally satisfying to pilots in a way they often could not easily explain, and which struck psychiatrists as vaguely neurotic.[55] "[F]lying higher and faster is the objective of most pilot types," astronaut Robert Crippen explained in a 2006 interview, summarizing the views of his colleagues.[56] For military aviators of the 1950s and early 1960s, flight-test work satisfied these professional needs until Project Mercury presented the pilots with the possibility of flying the highest-performance vehicle to date, as well as getting national attention, promotion, increased pay, and other benefits for themselves and their families.[57] Even those pilots dismissive of the Mercury spacecraft (like Schirra) were reluctant to see other colleagues take their place.

Had they not entered the Project Mercury selection, America's first astronauts would have likely pursued solid careers as test, instructor, and tactical pilots, as did the dozens of men NASA turned down in the 1959 selection.[58] Astronauts Michael Collins and Gene Cernan, in David Sington's 2007 documentary *In the Shadow of the Moon*, speculated with some ambivalence that they would have flown combat missions over Vietnam, as had many of their friends in military service. While pleased to have survived the period, Cernan resented having missed an opportunity to apply his training in combat alongside his colleagues.[59] Apollo 8 astronauts William Anders and Frank Borman, too, later reflected with some guilt upon on the relative safety of their NASA assignments and the wartime dangers they avoided through their participation in the space program.[60] Combat pilots were "over there, unsung, fighting a war that I believed could not be won," Apollo 15 astronaut Al Worden wrote in his memoir. "And what was I doing? I was sitting pretty in Houston, walking red carpets, designing pretty patches with a fashion designer. I always felt a bit funny about that."[61]

The professional attraction of Project Mercury was so great, despite its indignities, that the 1959 invitations drew candidates for the next several astronaut selections, and aviators like Pete Conrad and Jim Lovell who failed the first round of testing enthusiastically pursued later selections.[62] The interviews in Washington, DC included no medical examinations but did include several personality assessments. Two air force psychiatrists, George Ruff and Ed Levy, questioned applicants about their personal histories and challenged each to "sell" himself to his examiners.[63] Through successive waves of examinations, the initial applicant pool was narrowed to 32 applicants, who NASA subsequently invited to endure "exacting physical examinations" under great secrecy at the Lovelace Clinic, and psychological and "stress tolerance" testing at the Wright Air Development Aeromedical Laboratories at Wright-Patterson Air Force Base in Ohio.[64]

As active-duty military pilots, all of the Project Mercury applicants were in excellent health, and with the exception of the occasional individual with an acute medical issue, all were equally capable of functioning in space. (Indeed, many unsuccessful candidates enjoyed distinguished flying careers after the selections.[65]) Rather than weeding out the weak, examiners claimed the tests would identify those who, by virtue of their superior health and strength, might enjoy the longest spaceflight careers.[66] In fact, physicians and psychologists possessed no reliable data correlating their tests with professional success, and many of the tests had been invented just for the selections.[67] Some physiologists hoped that correlating the applicants' physical examinations with future performance data might enable more methodical astronaut selections in the future; at times, though, examinations at the Lovelace Clinic and at Wright-Patterson (where subsequent testing occurred) drifted into the realm of pure research.

Years later, astronaut Joe Allen speculated that the less extensive testing inflicted upon his colleagues in 1967 had merely allowed researchers to accumulate data on healthy people for projects unrelated to the selection.[68] In fact, aviation physicians Ulricht Luft, Paul Stapp, and others had conducted the Mercury evaluations mostly to exploit a hard-to-find group of test subjects for experiments intended to identify human tolerances for vibration, isolation, and extreme temperature. The data was scientifically valuable, but often yielded little information by which to distinguish one applicant from another.[69] Nude anthropometric photographs taken of the pilots to correlate "body typology" with flying ability proved particularly worthless, though; examiners only hoped that the "unique" data collected might "provide the basis for a long-term study " of the "relationship of man's physical constitution" to his "physiological performance in space."[70]

However bizarre, though, most physiological investigations did provide examiners with objective, if arbitrary, measures by which to compare seemingly identical individuals.[71] While the physicians ultimately ranked applicants by their performance on physiologic tests, it is unclear what role these rankings actually played in the final selection.[72] According to Scott Carpenter's memoir, the pilots' initial interviews in Washington DC had been the critical factor in their selection, and the medical examinations were merely perfunctory.[73] Indeed, the only physical qualification that was at all material to Project Mercury was a height restriction (candidates had to be less than 5' 11") required to ensure that the selected astronauts would fit inside the capsule.

Many pilots, according to Tom Wolfe, suspected that the true reason for the intrusive physical examinations at Lovelace was to verify both their psychological fitness and their determination to participate. One contemporary account of the Mercury selections confirms this, noting that the "main value of a severely stressful physiological test was the interpretation of the psychological response. . . . Whenever a subject terminated a severe test for psychological reasons, he was not recommended by the Committee."[74] For the applicants, the examination regimen Lovelace Clinic physicians imposed upon them was patently grotesque, which made the psychological dimensions of the testing all the more obvious. Indignities included the penetration of nearly every orifice of their bodies in ways that appeared to have been designed by experts to be as "humiliating" as possible. In one dreaded ordeal chronicled in the *Right Stuff*, pilots were led in open-back hospital gowns though the crowded clinic with balloons inflated in their rectums and rubber tubes snaking out of their anal cavities. To the pilots, the physicians demonstrated a particularly sadistic fascination with the gastrointestinal tract; prospective astronauts were asked to swallow a yard of rubber tubing and to submit to multiple enemas, and were held down by orderlies for painful rectal exams.[75] (Pete Conrad, Wolfe writes, openly objected to the repeated enemas and was subsequently rejected for his noncompliance.)

George Ruff and Ed Levy also conducted more conventional psychiatric examinations on the 32 finalists at Lovelace, aided by two psychologists enlisted to observe the applicants during testing.[76] Examinations included a battery of tests popular in the late 1950s, including the Rorschach Inkblot Tests and the Minnesota Multiphasic Personality Inventory, as well as a variety of standardized military aptitude exams for officer candidates and aviators, and stress tests to examine cognitive function under duress of isolation, noise, and other discomforts.[77] Summarizing an elaborate list of criteria, Ruff described the qualities he hoped to find in America's first astronauts, including generic characteristics like "intelligence," "drive," and "creativity." Other preferred characteristics seemed almost contradictory: independence combined with a willingness to "accept dependence on others"; ability to "respond predictably to foreseeable situations" and unforeseeable situations alike; motivation to succeed, but not out of a desire for "personal accomplishment"; and to be able to "tolerate stressful situations passively" while reacting quickly when needed, though never impulsively.[78]

Extrapolating their experience with previous laboratory volunteers (many of whom proved emotionally unstable), the psychiatrists feared the astronaut applicants would be thrill-seeking adolescents who used fast airplanes to assuage their sexual inadequacies. "In answer to the question, 'What kind of people volunteer to be fired into orbit?' one might expect strong intimations of psychopathology."[79] Indeed, as Ruff and Sheldon Korchin later wrote, the public tended to describe the astronauts as pathologically fearless.[80] Instead, Ruff and Levy found the group to be entirely free of "psychosis, clinically significant neurosis or personality disorder";[81] with the exception of a few applicants whose intellectual aptitude fell below minimum standards, no pilots were struck from the selection process on psychological grounds.[82] Desperate to locate people with the requisite technical skills, NASA had even determined to invite known eccentrics to interview,[83] but the psychiatrists instead found almost all of the examined applicants to be married, stable family men with excellent interpersonal skills and slight obsessive-compulsive tendencies.[84]

"Central in their personalities," Ruff and Korchin later wrote of these men, were "strong needs for achievement and mastery." Examiners suspected that many of the men had been, at some point in their lives, so tormented by feelings of inadequacy that they used achievement to "reduce self-doubts."

Eager for opportunities to prove themselves, the astronauts obsessed over their errors and exploited any opportunity to demonstrate their skills. Even in recreational activities, the astronauts competed ruthlessly, cheating on occasion to prevail against more skilled opponents. (Gordon Cooper had managed to beat Alan Shepard in illegal street races only because he had secretly modified his Corvette to ensure victory, a fact Cooper could not bring himself to admit until after Shepard's death.[85]) Failures, even for these seemingly taciturn men, were "keenly felt," but "they display striking resilience in the face of frustration," confronting disappointments with "renewed effort." "The fact is that emotions, both positive and negative, are strongly experienced," Ruff and Korchin noted of the men, "however, control is good."[86]

These controlled men were careful to hide, behind a veneer of sociability, personal failures and character quirks that might get them into trouble. While some astronauts (notably John H. Glenn Jr.) were likely as affable and mild-mannered as they appeared in their evaluations,[87] others either affirmatively misrepresented themselves or had been fortunate enough to have certain of their eccentricities excused by examiners faintly star struck by the pilots' impressive bearing and backgrounds.[88] Scott Carpenter managed to conceal his academic dismissal from college, and was given to occasional philosophical musings about spaceflight that concerned some NASA examiners. Gordon Cooper concealed his failing marriage, while Virgil I. "Gus" Grissom bore a barely concealed reputation as a foul-mouthed serial philanderer.[89] (Grissom once described himself to one reporter as interested chiefly in "flying and fucking," chagrined he could not do both at the same time.[90]) Peer evaluations conducted during the examinations occasionally revealed these faults: one unnamed finalist with a wicked sense of humor remarked that his colleagues were more promiscuous than they seemed.[91] "If you have to fuck your way into space," a certain pilot would likely be "the first man there."[92]

The pilots' attitude toward the experts who screened them was one of skepticism and mistrust; all thought of themselves as perfectly suited to space work and regarded NASA's bureaucracy as an impediment to achieving their goals. For test pilots, a modest yet confident appearance was a job requirement, and the would-be astronauts deployed it to mislead and impress their examiners. Seeking every possible edge in the competition, the men cooked the answers to screening questions, believing that they knew what responses would produce favorable responses from examiners. Ample social interaction between applicants, Tom Wolfe concluded (based upon the reminiscences of Pete Conrad), permitted applicants to trade information between interviews and coordinate their responses, making unique answers rare. Often the pilots were correct in their scheming; at other times, they merely managed to duplicate each other's mistakes. For example, when asked to "free-associate" on open-ended questions, pilots preferred to keep their remarks brief, fearing (excessively, as it turned out) that imaginative answers would mark them as disturbed.[93] (Conrad, a rare iconoclast, delighted in embarrassing the psychiatrists with patently bizarre remarks; when asked to free-associate about a blank sheet of paper, Conrad refused, insisting the sheet was upside-down.[94]) Similarly, when asked about their attitudes about risk, pilots also suspected (correctly, as examination reports reveal) that the psychiatrists were predisposed to regard them as suicidal. The pilots labored to contain their "daredevil" personalities, emphasizing their rational trust in the safety procedures common to test pilot work.[95] The examiners were satisfied with such answers, but would have preferred a greater acknowledgement of the danger. Nonetheless, Ruff and Levy alluded approvingly in their summary report to the pilots' measured, if overly optimistic, acceptance of risk.[96]

Like the stereotypical white collar workers of the 1950s, these would-be astronauts also seemed, to the examiners, somewhat dull. The pilots saw no monsters in their Rorschach Inkblot Tests; they saw nothing at all, seemingly lacking in "imagination and creativity." When it came to their motivations for participating in Project Mercury, spontaneous emotional responses were rare: pilots frequently cited engineering and technical interests, and admitted a mild enthusiasm for being on the forefront of technological innovation. Few would-be astronauts seemed at all interested in astronomy or exploration, ascribing their interest in spaceflight, to a desire for responsibility, career advancement, and new challenges within their existing line of work, like that of any budding business executive.[97] Most astronauts, psychiatrists found, were sensitive to the feelings of others but not inclined to cultivate them; they were variously "effective" in "dealing with people," but avoided deep relationships.[98] The pilots' aloofness and seeming lack of understanding of their own motivations concerned the psychiatrists, as did their relative insensitivity to danger, but the examiners were, overall, impressed by this confident, capable group of flyers.[99]

By concealing their personalities and overcompensating for traits they assumed would disqualify them, the would-be astronauts emerged to their examining psychiatrists as slightly robotic. Though disappointed that so few appeared to be introspective or self-aware, the examiners did not reject applicants on this basis, especially as such traits did not materially impact their professional abilities, and as the traits seemed uniformly distributed among the pool of applicants. The early astronauts may not have been the dullards Ruff and Levy described, but they were ambitious and smart enough to suppress adolescent enthusiasms to get along in a high-profile crash program that had definite ideas about what it wanted in its spacemen. Ironically, astronauts who fully met Ruff and Levy's description might indeed have been those best suited to Project Mercury; instead, NASA secured the services of men who secretly harbored ambitions and enthusiasms about which they had not been completely forthcoming.

Final deliberations on astronaut selection occurred in March 1959, and Gilruth approved the preliminary selections on April 1. Despite the copious data collection, it is unclear on what basis the final selection decisions were made or by whom. George Low may or may not have been present in the deliberations, while Lovelace, Flickinger, and civilian test pilot

Scott Crossfield are known to have participated.[100] Primary responsibility, though, appears to have fallen to three members of the original Selection Committee: Project Mercury assistant director Charles "Charlie" Donlan, test pilot and Space Flight Programs chief Warren North, and psychologist Allen Gamble.[101] Before proceeding further, the Committee members immediately ruled out eight candidates they described as possessing "character traits undesirable in the team effort."[102] With 23 candidates remaining, though, the committee members found themselves unable to make the final cut; seeking only six astronauts, they eventually settled on seven after Gilruth ended the stalemate and invited the committee members to recommend an additional candidate.[103]

Examiners could not help describing the men selected—Gordon Cooper, Gus Grissom, and Deke Slayton of the air force; John Glenn of the marines; Scott Carpenter, Alan Shepard, and Wally Schirra of the navy—as vaguely Victorian and stereotypically American. They had emerged from "middle-middle" or "middle-upper" class families and, as oldest or only children, had endured few childhood traumas or setbacks.[104] Raised in small towns, they could recount stories of almost continuous achievement in public high schools and state universities. All were described as self-effacing, good to their parents, responsible, and decent, and all were Protestants, though religious observance among the men varied.[105] Four of seven selected astronauts, psychiatrists noted with wonder, were named "Jr."

The men filling America's earliest astronaut ranks were so often first-born, eldest sons, or only children that psychologists and journalists could not help but notice the pattern. A 1968 *New York Times* article wondered why all three members of the *Apollo 8* crew, the first to circumnavigate the Moon, were only children.[106] The coincidence so interested Daniel Patrick Moynihan, Assistant to the President and author of an influential study on the Negro family, that he requested NASA investigate the coincidence.[107] A report by NASA's Dr. E. J. McLaughlin dated June 2, 1969, and forwarded to Moynihan on June 10, confirmed that 87.5 percent of all astronauts in active service at the time were only children or eldest sons; while, of the 23 astronauts who had flown in space, 21 fell into this category. Even the two flown astronauts who were not eldest sons were raised as if they were, due to the early death of an older sibling in one case, and the large age disparity between siblings in the second.[108]

NASA psychiatrists never felt entirely comfortable with their role in selecting the first astronauts, and George Ruff was left without "any feeling that the psychiatric screening itself had been all that important." NASA insisted that psychiatric examiners only "screen-out" mental illness in applicants, and were unsympathetic to their proposals to identify persons with qualities that might render them particularly well-suited to spaceflight careers. Absent testing procedures customized to would-be astronauts, the psychological evaluators substituted basic mental health testing, evaluations of applicants' "social skills," past military performance reviews, and peer evaluations for more elaborate testing, eventually recommending virtually the same applicants that other examiners favored. In part, NASA's reluctance to expand psychology research was explained by the lack of any clear understanding of what stresses space work would actually entail. Due to the lack of data (and NASA's reluctance to generate more), the summaries created by Korchin, Levy, and Ruff would prove the only formal personality profile for astronauts until 1987.[109]

Enter the Astronauts

As later described by Tom Wolfe, the early astronauts were antiheroes: simple, natural pilots who flew for themselves, regarded authority with suspicion, and bucked the patriotic culture from which they emerged. This image, though, was an artifact of the 1970s, the decade in which *Right Stuff* was composed. At the time of their selection, NASA pitched the astronauts as exactly the kind of loyal, modest men America needed to confront the Soviet Union in space. *Sputnik* had presented Americans with a sinister superstate seemingly able to organize high technology better than its Western counterpart. In America's would-be astronauts, NASA would offer a human face: characteristically American figures—honest, energetic, reverent—who would master space technology and claim America's rightful place in the heavens.

NASA introduced America's first seven astronauts on April 2, 1959, and presented them to the public on April 9, at a press conference convened at NASA Headquarters in Washington, DC's Dolly Madison House by Walter Bonney, director of NASA's Office of Public Information. Reporters and photographers could barely contain their enthusiasm, describing the selected astronauts as the results of a "long and unprecedented series of evaluations."[110] First to speak were not the seven astronauts, but the individuals who defined the persona of NASA's astronauts and had chosen the men to fill it—T. Keith Glennan, General Don Flickinger, Robert Gilruth, Captain Norman Barr (director of the navy's Astronautical Division of the Bureau of Medicine and Surgery), Charlie Donlan, and Randolph Lovelace— "highly intelligent" and "highly motivated" "family men" notable for their "stability" and "powers of observation." NASA hoped its clean-cut pilots would stimulate public confidence in the space program and trumpeted the men as courageous soldiers, loyal public servants, and scholars, posing them in jackets and ties before the press. "These men have been chosen from a population of about 180 million," Barr declared, "to represent the United States in this important project. We are all behind them a hundred percent."[111]

As later described by Wolfe, the April 9 press conference was a circus in which sympathetic reporters peppered the new astronauts with softball questions and eagerly absorbed their often inane answers.[112] While the transcript bears much of this out, though, the journalists' enthusiasm for the astronauts was tempered by their impatience. Reporters badgered NASA managers for concrete information about launches, while NASA managers deflected questioning by touting their photogenic but untested astronauts. While some astronauts eagerly pandered to managers and reporters, others slowly began to shed their accommodating demeanor and to define their role in the space program.

In response to early biographical questioning, the astronauts were stiff and taciturn; asked about the support each received from his "good lady" and "children," the astronauts avoided detailed answers, describing themselves as men who made their own professional decisions. Only Glenn adapted quickly to the demands of interviewers and flattered their preconceptions, describing how his loving family supported his dangerous, lonely work "one hundred percent." When asked about their motivations for joining the space program, again, all but Glenn kept their answers vague and formulaic, describing their desire to serve their country and to get in on the ground floor of the latest innovation in aviation. "We are interested in new things," Schirra declared. "I am a career officer, career pilot," declared Cooper, "and this is something new and interesting." None mentioned scientific curiosity.[113] When one questioner aggressively solicited the astronaut's religious views, Glenn again attempted a long answer that appeared to satisfy the interviewer's expectations, describing his Sunday school teaching and desire to get as close to "Heaven" as he could.

In the 1983 film adaptation of *The Right Stuff*, Shepard greedily eyes Glenn's success with reporters and parrots his piety with an oily grin; in the press conference, though, Shepard was among several astronauts who eloquently disagreed with the journalists. Schirra, Shepard, and Slayton responded aggressively to the reporters' religious inquisition, dismissing the questioning and attempting to correct what they described as public misunderstanding of the nature of space work. Displeased by the reporter's suggestion that only people with strong religious convictions would pursue test piloting, Schirra told of his faith in "mechanical objects" and the "machine age," while Shepard, though trying not to "slight the religious angle," insisted that the hazards entailed in Project Mercury had been exaggerated, a point Slayton reiterated. As professional pilots in a "professional program," the men declared, they thought little of God, and would take comfort only in their own technical abilities and the skills of those with whom they worked.[114]

Confronted by public expectations, the astronauts responded differently: some astronauts, like Glenn, accepted the public nature of their work and acknowledged the need for self-presentation, exploiting new forms of media communication (including television) to build public confidence in the space program. Cooper, meanwhile, clung to earlier ideals of the aviator as an individualist and adventurer; Grissom turned inward, embracing an austere engineering ethos; while Carpenter embraced the freedom of thought associated with the incipient counterculture, to the chagrin of colleagues. Schirra, Shepard, and Slayton, though, recognized their role as technical experts in a new techno-scientific infrastructure, and sought to exploit their skills, credentials, and celebrity to wrest control of it from the civilian engineer-managers who would soon multiply in NASA. While certain journalists preferred Glenn's formulation of the astronaut as a pious, devoted patriot, it would be Schirra's project management concerns that would dominate as the astronauts attempted to define their role in the space program.[115]

Journalists at the press conference seemed unsatisfied with the astronauts' responses, demanding more specific information on when the men would fly and in what order, questions to which Gilruth had no answers. Except for Glenn, the pilots stubbornly resisted the heroic persona imposed upon them, and toward the end of the conference reporters began to ignore the astronauts completely, pestering NASA managers about upcoming launch plans. The reporters' persistence lead Bonney, at one point, to ask the journalists to "stick to questions about the men themselves."[116] Ironically, NASA, which had once sought to anonymize the astronauts, now needed their celebrity to protect the space program from allegations that it was a skeletal effort with no plans to challenge the Soviets in space. The astronauts, meanwhile, labored to counter impressions that they were reverent public servants with little to say.

Reporters commenting on the proceedings appeared to miss many of the subtleties. Media reports satirized the "dramatic flourish" of the astronauts' debut but bought into NASA's description of the men as courageous and representative.[117] These "plain-speaking small town fliers," wrote James Reston after the press conference, seemed so confident and free of cynicism as to be inspiring even in their insipidness. "Somehow they had managed to survive the imagined terrors of our affluent society," wrote Reston, mocking the latest sociological treatises, "our waist-high culture, our hidden persuaders, power elite and organization men."[118] The astronauts, though, were not frontiersmen or farmers, but the sons of America's professional and managerial classes.[119] Indeed, 66.7 percent of the fathers of America's astronauts of the 1960s, according to E. J. McLaughlin's 1969 NASA study, were "professional," "technical," or "managerial" workers (including military officers), while another 19.4 percent were engaged in "clerical and sales" work.[120] Farmers and tradesmen were far more rare among the fathers of the astronauts than they were among the general population, or even among NASA's scientists and engineers, most of whom hailed from blue-collar households.[121] Instead of homesteaders, NASA had secured the services of skilled, highly educated men from professional households who, like junior executives all over the United States, had earned the attention of their prospective employer—NASA—because they possessed exclusive educational credentials and spotless service records. Astronauts' disputes with NASA's engineering community did not, as astronauts sometimes suggested, pit salt-of-the-earth pilots against privileged NASA intellectuals, but the other way around.

The astronauts, moreover, had triumphed by fitting in. If some journalists conceived of astronauts' professional accomplishments in terms of William Whyte's inner-directed "Protestant ethic," their successes would have been impossible without the "social ethic" of contemporary businessmen. Coworkers like NASA nurse Dee O'Hara remembered many of the astronauts as smooth charmers; ingratiating and eager to please.[122] When no one was around to impress, though, some could be very different people. Grissom, Slayton, and Shepard in particular, were hardly the gregarious men they appeared to NASA management or the public, and were often aloof, shy, and even cold in private. Alan Shepard could be both friendly and extremely cruel, interrogating underlings until they became flustered.[123] This treatment so irritated O'Hara that on one occasion she shouted at Shepard for teasing her. To her surprise, Shepard relented; O'Hara concluded that Shepard secretly craved the approval even of those he berated.[124] Most astronauts soon learned how to moderate their image for the press; aware that their personalities could grate upon others, the men thrived during the selection process by cleverly altering their personalities to suit whoever happened to be in the room. NASA Public Affairs chief Paul Haney described this talent as the ability to "dial up the emotional commodity that's needed" to suit whatever social situation had been thrust upon them.[125]

"What Does an Astronaut Do?"

Despite NASA's efforts to control their exposure, astronauts became NASA's public face, and used that attention to manipulate their workplace in the same way that they had manipulated the examiners who had poked, prodded, and interrogated them.[126] Though instantly famous, NASA astronauts found themselves occupying a professional middle ground between two camps doubtful of the contributions they could make. Wary of entrusting such complex machines to human control, NASA engineers sought to place the astronauts in automated capsules designed for expendable lower-order primates. Meanwhile, experienced test pilots regarded the astronauts as sell-outs who were wasting their talents—"spam in the can" tasked merely with staying alive in an expensive metal box.[127] In order to satisfy doubters, the astronauts would need to become something more than mere lab specimens, mollifying the test pilot community and asserting their dominance over NASA's engineers and physicians.

This would not be easy. The idea that NASA had recruited, in the form of its first astronauts, the best aviators that the nation could produce was an exaggeration; none of the astronauts of the 1960s were counted among the nation's most respected pilots, and some, astronaut Walter Cunningham later asserted, were so sloppy in the cockpit of fighter jets that other astronauts privately avoided flying with them.[128] Most were junior officers in their early thirties: promising young officers but not established leaders in their fields; of the "Original Seven" only Deke Slayton had served in the elite Fighter Branch of Test Operations at Edwards, the *crème de la crème* of this small crop of test pilots.[129] At the time of their selection, only two of NASA's seven astronauts—Glenn and Slayton—were SETP members, and selection for Project Mercury did not automatically qualify the others for membership.[130] Once at the space agency, astronauts worked with engineering teams to design space vehicles and associated hardware, and made substantial technical contributions. Yet, against the backdrop of skepticism from their test pilot colleagues, many of NASA's early astronauts reveled in the conceit that they were not professional rocket men, but elite aviators, reluctantly drafted into space service.[131]

While astronauts a decade later would position themselves as intuitive aviators undermined by book-trained scientists, what originally distinguished the astronauts from the rest of the nation's best pilots were their organizational talents and technical skills. Writing in 1966, Kenneth Keniston seized upon the technical expertise of the astronauts, and not their flying skills or supposed "heroism," as their defining trait. Preternatural ability or heedless independence did not make astronauts who they were, exacting professionalism did: engineering scholarship, laborious flight training, and skill accumulation that earned them the respect of colleagues, managers, and the public.[132] Schirra, for example, spoke of the "hot shot pilot" as "not just a scarf and goggles type, but one who could use his engineering confidence to work on systems and make the best airplane, ever."[133] Armstrong, a civilian test pilot for NACA before joining NASA in 1962, distinguished himself at Edwards, James Hansen recounts, more for his understanding of the mechanics of flight than his piloting. To Yeager and other experienced test pilots, Armstrong was not the kind of intuitive pilot they preferred in the cockpit, but an accident-prone junior naval aviator who lacked the instincts to avoid in-flight mishaps, but whose studious demeanor and academic training permitted him to recover skillfully from them.[134] It was Armstrong's enthusiasm for engineering that ultimately endeared NASA to him.[135]

Project Mercury would offer these ambitious pilots very little opportunity to actually fly their vehicles, a fact that NASA psychiatrists George Ruff and Ed Levy did not seem to fully recognize. Their summary described Project Mercury only as a "two-year training program, followed by a series of ballistic and orbital flights" of uncertain duration and with little or no piloting responsibility. *What Does an Astronaut Do?*, a 1961 youth nonfiction book by Robert Wells, described the spacemen's role perhaps most accurately:

> The astronaut sees to it that his spacecraft does the job assigned to it. Its control system for oxygen and air pressure, its control system which keeps the craft "right side up" in flight—and most of its other systems—can work automatically. They usually do. Yet the astronaut is in command of these systems and their controls.
>
> He is manager of all these systems.[136]

NASA, effectively, needed not pilots, but officer-managers, who would supplement mechanical feedback and control systems with their own senses, monitoring, supervising, and, only if necessary, controlling their spacecraft. Yet, while noting that "the pilot's duties will consist largely of reading instruments and recording observations," Ruff and Levy nonetheless accepted the assertion of most applicants that participation in Project Mercury would be enjoyable piloting work, would enhance their flying careers, and represented a natural step in the evolution of aviation.[137] In fact, some of the astronauts had deep misgivings about the role of piloting in the Mercury spacecraft; wary of criticizing the program during their interview, the would-be astronauts simply lied.[138]

Addressing the annual SETP convention in October 1959, though, Slayton had put aside his notoriously artless speaking manner and skillfully disarmed a room full of test pilot critics in a presentation entitled "Operational Plan and Pilot Aspects of Project Mercury," which suggested the astronauts' future strategy in defining their professional role.[139] Yes, the first astronauts would follow monkeys into space, and possibly do little more than monitor automated systems and respond to emergencies. If humans were to have any piloting presence in space, though, Slayton suggested that this modest role was a necessary step dictated by technical and biomedical uncertainties; condemning Mercury would only slow the evolution of flight into space.[140] Mercury might be humiliating, but it was necessary, and rather than dismissing it, potential astronauts needed to recognize both their limited operational control of the vehicles and the substantial influence they still possessed over the future course of spaceflight.[141] While astronauts would work to consolidate their authority against the wishes of engineers and physicians, they would also acknowledge a relationship with their craft subtly different from that of the conventional test pilot. No longer steering their vehicles directly, the astronauts could manage the program that built, operated, and promoted them and mediate between the engineering and flight test communities.

The targets of the astronauts' offensive were the engineers building their capsules and the physicians who had selected the astronauts to fly in them, both of whom the astronauts regarded as novices to the world of experimental flight. The early capsule designs that NASA's aerospace engineers had produced lacked critical safety features—backup systems, manual overrides, hatches—that test pilots expected to find on experimental vehicles.[142] Automated and remote-controlled systems pleased NASA engineers unwilling to subject hardware to the vagaries of human performance, but at a time when such systems were still in their infancy, pilots regarded them as supplements to human control, not replacements. A vehicle in which the human pilot retained override capability, astronauts argued, offered greater chances for mission success, as well as greater mission flexibility.[143] Just prior to the formation of NASA, NACA research had determined that "proper use of the pilot may considerably reduce the complexity of the vehicle and increase overall reliability, particularly in case of malfunctions."[144]

The astronauts took this responsibly seriously, and were not alone in their efforts; with the support of NASA managers like Robert Gilruth and his deputy, respected engineer Max Faget (who worked diligently, for example, to develop Mercury's escape systems), the astronauts insinuated themselves into vehicle design and mission planning.[145] In *We Seven*, Slayton recounted how the men, upon their arrival in NASA, had divided various engineering aspects of Project Mercury among them—Carpenter to navigation, Cooper to the Redstone, Glenn to cockpit design, Grissom to flight controls, Schirra to spacesuit development, Shepard to recovery systems, Slayton to the Atlas—and asserted themselves in design work (Figure 1.2). Astronauts took to these assignments with alacrity and enthusiasm, exploiting their notoriety to push

Figure 1.2 Astronaut John Glenn completes a training exercise in the Mercury Procedures Trainer at Langley Field, Virginia (NASA photo)

design changes. Grissom, in particular, was so obsessed with the design of the capsule's autopilot, Slayton recalled, that one NASA manager eventually exclaimed that if the device did not work, it would be Grissom's fault.[146]

Interactions between astronauts and ground engineers were occasionally chilly, but always productive. "[W]e had a fair amount of prestige around the country," noted Wally Schirra at the time, "and though we did not always succeed in getting what we wanted, we sometimes ganged up—all seven of us together—and the extra weight helped us to win at least a compromise." Astronauts valued robust design, redundancy, and flexible operation; engineers favored precision systems making efficient use of volume and weight allowances. (Engineers took to calling the spacecraft's many backup systems as "redundant" equipment, but astronauts took offense even at this term: anything that kept them alive was a vital system.)[147] Often, the proposed design changes—like the addition of a window on the Mercury capsule—were remedied with little debate.

To some NASA engineers, the astronauts' insistence on redesigning capsule hardware to suit their tastes may have been infuriating, but the astronauts' participation in Mercury engineering work never rose to the level of mutiny.[148] Indeed, the astronauts' independence was a professional mindset borrowed from military flight test programs and encouraged by STG managers who were disposed to trust test pilots and wanted to exploit their skills and experience. Gilruth, in particular, encouraged the astronauts to push for redesign of equipment they felt unsuitable, and later recalled the men as doing excellent work.[149] "People used to tell me that I had no control over the astronauts," Gilruth later remarked, "I'll tell you those boys were wonderful."[150] As the ones ultimately responsible for flying the vehicles, the astronauts argued they possessed the best idea of what would—and would not—work in space; their status and the support they enjoyed from Gilruth thus made the astronauts vital arbiters in the design process. "[W]e decided that since we were the test pilots who would be flying the thing, we had a right to stir things up a bit," Slayton recalled. "That is what they hired us for."[151]

The astronauts' authority was more circumscribed in contests with NASA's flight surgeons, who had the authority to pull them from duty for a host of medical reasons.[152] As pilots, the astronauts knew that physical injury or illness could derail their careers, and their community proved so reluctant to seek medical attention that NASA assigned Dee O'Hara to privately monitor them, in the hope that the astronauts would confide in her. Astronauts recognized that however approachable the health care provider, physical infirmity disqualified them from flight, and they resisted medical oversight: Grissom, fearing he had fractured his left wrist in the weeks before his 1965 Gemini 3 flight, attempted to treat it in secret, fearing he would be replaced.[153] As a group, the doctors proved more challenging to undermine than did the engineers; they eventually grounded several of the Original Seven for medical reasons. Even worse, the physiologists' enthusiasm for primate testing threatened to undermine the timetable for piloted flight.

Despite their efforts, as early as April 1959, it was virtually certain that none of the seven astronauts would actually become the first living creatures, or even the first Americans in space. A month after orbiting the first artificial Earth satellite in 1957, the Soviet Union had orbited a dog, Laika, who died of heat exhaustion days—or possibly even hours—into her seven-day flight and was not recovered.[154] The Soviet Union had been conducting less ambitious dog flights since 1951; suborbital primate flights, to which Chuck Yeager had referred in mocking the Mercury astronauts, had been a staple of US rocket research since 1948, when the army placed monkeys in the nose cones of captured German V-2 missiles. On May 28, 1959, army engineers launched two monkeys aboard a Jupiter intermediate range ballistic missile to an altitude of 360 miles, experiencing nine minutes of weightlessness until they fell back to Earth on a suborbital trajectory and were recovered. As Yeager had predicted, NASA tested its Mercury-Redstone and Mercury-Atlas hardware with apes before humans, launching chimpanzee Ham on a suborbital flight in January 1961, and orbiting the chimp Enos the following November. While Scott Carpenter later blamed NASA Headquarters for delaying piloted flights to gain redundant biophysics data for the medical community, primate flights served to test both life support systems and launch vehicle hardware.[155] Ham the chimp, on his January 1961 suborbital flight, splashed down off-course by more than one hundred miles due to a hardware problem, while an exploding air force Atlas "E" ICBM killed a squirrel monkey shortly after its November 1961 launch.

The astronauts were opposed to primate flights not, as some suggest, because monkeys were more competent in the cockpit, but because they delayed the flight schedule, and because a remote-controlled or automated vehicle that apes could fly was more apt to fail catastrophically.[156] Such a craft also held no professional interest for trained pilots, who had always viewed cockpit automation with measured, professional skepticism.[157] Air force literature of the period was explicit: while the presence of certain navigational and computational devices in the spacecraft cockpit was inevitable, any device to fully replicate the human brain would likely weigh a thousand times as much and be impossible to mass-produce.[158] The astronaut's job was not that of pure hands-on pilot, but nor would it be mere passenger: only a partially automated vehicle could complete necessary orbital maneuvers, but it would likely fail catastrophically without constant human monitoring.

As much as astronauts claimed to resent electronic computers, though, astronauts soon realized that they could not fly without them: "the speeds and distances involved...made the pilot's eyeball useless," astronaut Michael Collins later wrote, "gimmicks such as radar and computers have to be carried on board," supplemented by "a whole bevy of geniuses on the ground... with their own, more powerful, radars and computers."[159] The new avionics hardware, though, would inform human decision-making, not replace it, making human pilots more self-reliant by obviating the need for ground control during critical maneuvers.[160]

Often astronauts banded together to challenge NASA superiors on design matters, but the men, critically, never regarded each other as equals, and remained divided. "If they were astronauts, they were men who worked for the team," Normal Mailer later wrote of the ones he observed in 1969, "but no man had become an astronaut who was not sufficiently exceptional to suspect at times that he might be the best of all."[161] Though all skilled military test pilots, some astronauts had enjoyed more flight test responsibility than others: among the former air force test pilots at Edwards, according to Gordon Cooper, operational duties distinguished "engineering" test pilots and more elite "flight" test pilots who supposedly undertook more dangerous and difficult flying work. Even as old job titles evaporated in Project Mercury, Cooper recounted with some residual bitterness, former Edwards flight test pilot Deke Slayton dismissed the abilities of former engineering test pilots like Cooper, though both men possessed nearly identical skill sets, educational backgrounds, and military pedigrees.[162]

Man in Space

For the astronauts, their arrival in the NASA meant the beginning of the real competition: a contest among seeming equals for the first piloted flight. "[E]valuations of each member of the group are almost identical," Korchin and Ruff noted in 1964, "except for the issue of who best qualified for a flight."[163] The quest for prime crew assignments became the defining effort of the astronauts' professional lives; while astronauts later enjoyed input in assignments, at first NASA managers chose crews and jealously guarded their decision-making process. At the April 1959 press conference that introduced the astronauts, Walter Bonney announced that NASA had not yet determined which man would fly first, and that even more surprisingly, "[h]e won't know himself until the day of the flight."[164]

In early 1961, Robert Gilruth asked the Mercury Seven to rate each other's abilities by suggesting in what order they should fly, and why. After consulting with Charlie Donlan and Walter "Walt" Williams, Gilruth determined the flight order, announcing it to the astronauts privately in a small classroom.[165] Shepard had won the coveted first flight; Grissom received the second flight. Schirra and Glenn appear to have been particularly galled: Glenn, that he had been placed third, and Schirra, that NASA had waited so long to tell him that he had been relegated to one of the last flights on the schedule. According to Shepard, the deflation in the air at the news was palpable; all of the astronauts believed themselves capable of performing the mission best, and many were miffed by how little say they had in the process.[166] "For most of the men the hardest problem to master was not being chosen for the first flight," Ruff and Korchin later reflected, "[t]he immediate disturbing implication was that those not selected were not doing as well as they had thought."[167] "Quite frankly," Schirra described in his 1988 memoir, "the way the flight selection in Mercury was handled by Bob Gilruth and company baffled and disturbed me."[168]

This first assignment experience was so unpleasant for the astronauts that it was the last time any would permit assignment decisions to be made in this way. Later, the astronauts conspired to create an internal leadership structure that could take assignment decisions out of the hands of civilians. Until then, though, Gilruth's STG, an "agency of a bureaucracy," was too powerful to challenge directly, so the astronauts informally canvassed colleagues to restore some measure of control over their careers.[169] According to Schirra, Glenn soon began to lobby his colleagues to have his third, suborbital flight converted into an orbital mission; Shepard would later make a similar appeal for a second Mercury flight.

Though the chimpanzee Ham's flight suggested that NASA could successfully launch and recover an astronaut, an additional unpiloted Mercury-Redstone flight in March 1961 robbed Shepard of the opportunity to become the first human in space.[170] Having already lost the space race to dogs and chimps, America's astronauts were further disappointed in April when the Soviet Union announced that human pilot Yuri Gagarin had made a single orbit of Earth aboard his Vostok spacecraft. While America had cultivated a new kind of white collar space professional, the Soviets had—yet again—achieved a major space first with surprisingly simple equipment and unsophisticated personnel.

As air force and NASA managers debated selection criteria for astronaut candidates in the late 1950s, in the Soviet Union a similar controversy had raged between two conceptions of the space traveler, with advocates of a less independent pilot role eventually prevailing. While Soviet Air Force

Lieutenant General Nikolai Kamanin (supervising cosmonaut selection and training) advocated for a more direct control of the spacecraft, master designer Sergei Korolev favored an automated capsule with a redundant human operator tasked only to survive. Korolev's view proved more influential; while American astronauts established themselves as a bulwark against what they perceived as shoddy engineering, cosmonauts dutifully submitted to a program that regarded them as unlikely to make meaningful contributions to either vehicle design or flight testing. For American astronauts, psychological screening ended with selection; from then on, the astronauts slipped effortlessly into the role of engineering test pilots. For cosmonauts, psychological examination and conditioning remained a central aspect of their working lives, with training consisting largely of rote repetition, extreme sensory deprivation, and intense physical training, intended to induce mechanical perfection in the performance of ground instructions, even under circumstances of extreme stress.[171] NASA fretted about safety and sought white-collar professionals in its capsules; the Soviet space program wanted heroes of socialist labor.

Gagarin's flight epitomized the combination of audacity, paranoia, and social control that characterized much of the Soviet space program to follow. Young, polite, and agreeable, Gagarin, at age 25, had barely finished flight school when the Soviet Union secretly tapped him and 19 other compact, pliable military pilots for its human spaceflight program. The Vostok capsule, intended to carry humans into space but sold to the Soviet military as an automated reconnaissance platform, was voluminous but rudimentary, with a reentry system that frequently failed. Assigned to occupy his spacecraft only days before his flight, Gagarin at no point took control of it. Unsure whether he would survive impact on the frozen steppe, Korolev had ordered Gagarin to eject from his craft during descent and return to the ground by parachute. As international consensus required aviation record holders to return to Earth in their vehicles, the Soviets had concealed the details of Gagarin's return from the world press, as it did with virtually all information about the flight. TASS, the Soviet news agency, announced Gagarin's flight only upon his return, releasing little information about Vostok or the launch vehicle that had carried it.[172]

The simplicity of Soviet equipment and the Soviets' lack of openness of the concerning spaceflight activities (as well as a history of accidents among its cosmonaut-trainees) fueled repeated rumors on both sides of the Atlantic that Gagarin had not been the first Soviet citizen launched in space, only the first to return alive. Soviet secrecy also provided the United States with an opportunity to distinguish its relatively open program with the obviously military-controlled, highly secretive Soviet space endeavor. If the United States could not compete effectively with Soviet power, policymakers asserted, at least it would do so openly, in full view of the world, dedicating the scientific knowledge it produced to the public trust. Soviet secrecy, though, concealed the fact that its lead was not as great as US intelligence experts feared: Soviet achievements had been purchased at the cost of a well-organized developmental program. With stronger encouragement and less interference from the Kennedy administration, Gilruth later speculated, NASA might have beaten the Soviet Union to the first piloted space flight. Laborious testing of boosters and subsystems, as well as primate flights, though, were fundamental to the American design process.[173] NASA, John Glenn asserted, wished to be "right, not first."[174]

To Alan Shepard, regarded by many in NASA as the most capable of the Original Seven, fell the responsibility for the first Mercury suborbital flight in May 1961, but coming on the heels of a chimp flight and just before later orbital missions, even Glenn could not resist referring to Shepard as the "missing link" between ape and man. (Scientists at the Massachusetts Institute of Technology dismissive of human spaceflight were even less kind, referring to him, according to Gilruth, as "after the chimp, the chump."[175]) If America's answer to Gagarin's flight did not quite match Soviet achievements, it did suggest how Americans would define success in missions to come. Not only did Shepard complete an elaborate series of tests and experiments during his five minutes of weightlessness, he rotated his spacecraft using the manual control stick for which the astronauts had argued so effectively. In flight, Shepard avoided idle chatter and spurious reflection upon the environment of space, eschewing tourism for the checklists of a thorough flight test program. After returning to Earth in his capsule by parachute and splashing down in the Atlantic, Shepard, once aboard a recovery ship recorded a second-by-second account of the mission so "lucid" that examining physicians were dumbfounded.

What quickly characterized the best performances in space was precision, and what damaged the reputation of astronauts was any intimation that they had panicked, a charge leveled, probably unfairly, on Grissom after his 1961 flight ended with a hasty helicopter rescue and the spacecraft sinking to the bottom of the ocean.[176] John Glenn's long-awaited orbital flight in February 1962 was marred by an instrumentation problem that suggested the capsule's vital heat shield had partially detached, leading to Glenn's possible vaporization when the spacecraft reentered the atmosphere at high speed. NASA flight director Chris Kraft ordered Glenn's flight cut short after only three of the planned seven orbits, but Glenn's demeanor during the flight was unaffected by the near-catastrophe and he landed successfully, to a hero's welcome.

Wally Schirra's flight in October 1962, the third American orbital mission, is similarly remembered principally for its crisp exactitude. On his flight, Schirra performed in a fashion that, as Shepard and Slayton recalled, would have made a "robot" jealous, executing maneuvers with a deft touch and concluding the mission with enough thruster fuel left over to fly the mission all over again. Schirra landed only four-and-a-half miles from an awaiting aircraft carrier, an astonishing feat of accuracy. Schirra claimed that it was the recovery ship that was out of position and not he, but in either case, the performance was astonishing.[177] Gordon Cooper's May 1963 flight not only placed him in space for more than a full day, but ended with an even more precise landing, in spite of the failure of all of his automatic systems.

The New Astronaut Order

By 1963, Mercury had proven an obvious success and work was already underway on an enlarged vehicle, Gemini, carrying a second crew member and capable of various operations—orbital maneuvering, rendezvous, docking, spacewalking, and long-duration flight—that NASA needed to perfect to complete President John F. Kennedy's 1961 challenge to "land a man on the Moon and return him safely to Earth" by 1970. Cooper's 1963 Mercury flight was the last; having logged only 15 minutes in space in 1961, Shepard had lobbied vigorously for a tenth Mercury-Atlas flight that would put him in space for three days or more; eager to move on to the two-person Project Gemini flights, NASA administrator James Webb was reluctant to invest in further Mercury missions. Even with the support of Gilruth and Walt Williams, Shepard was unable to change Webb's mind, and appealed personally to an unsympathetic President Kennedy to intervene on his behalf.[178] Newspaper coverage of this rare act of "insubordination" delighted journalists with the idea of labor unrest among the astronauts; one account described how their "lobbying" efforts and threats to " 'sound off' publicly" were viewed with "disfavor and annoyance by some high officials in the Administration." In attempting to set policy, though, the astronauts had overreached their authority, and Webb and NASA associate administrator Robert Seamans quickly silenced them, announcing their decision without entertaining debate with the astronauts on the subject.[179]

The controversy over Shepard's abortive second flight eventually established the upper bound of the astronauts' authority: they could manipulate hardware and influence training, but they had no power to alter strategy, except through success or failure. While Shepard attempted to solicit political support for an additional flight, Wally Schirra, with similar difficulty, demanded more authority over his missions: he saw his role as that of a "commanding officer" directing operations. NASA, though, would not allow astronauts that kind of control. "I was trained to be a commanding officer," Schirra later recalled, "and NASA never understood what a commanding officer was."[180] Rather, NASA co-opted the astronauts' professional talents, encouraging them to improve the program, but only as members of a team, and within the strategic framework established by NASA's civilian leadership.

The subsequent medical disqualification of Deke Slayton and Alan Shepard, though, eventually consolidated the astronauts' control over the few aspects of the spaceflight program that they could actively manipulate. In 1962, NASA grounded Slayton for an idiopathic cardiac arrhythmia on the eve of what would have been America's second orbital flight. Scott Carpenter flew in Slayton's place, and Slayton was forbidden even to fly jet aircraft solo.[181] Already a public figure, though, Slayton could not easily be jettisoned from the space program; instead, Robert Gilruth, at the request of the remaining astronauts, created an administrative position to occupy him. For Schirra, Slayton's appointment both satisfied his colleagues' sense of duty to a grounded peer and insulated them from the possibility that NASA might appoint an external administrator to manage the astronauts, as was rumored might occur. "Deke had been through hell. But we were proposing him as our leader out of respect not pity."[182] Meanwhile, the remaining astronauts carved up future Project Gemini flights for themselves.

Sidelined to a supervisory position so as to preserve his reputation and the public image of the space program, Slayton could have easily avoided responsibility. Instead, he pursued the management of his fellow flyers with extraordinary diligence. The transition of the Astronaut Office from external management, to consensus, to internal control was a complex one; with the Astronaut Office's move to the Manned Spacecraft Center in Houston Texas in 1963, flight rosters became a complex organizational problem managed by Slayton. As an administrator, Slayton eventually proved capable, and controversial: his inability to fly lent him an air of impartiality that made his decisions appear to be beyond reproach. Though hoping one day to return to flight, Slayton no longer competed with colleagues for missions, and was in a unique position to evaluate colleagues and mediate their disputes over flight assignments. His pedigree—an engineering degree, an air force commission, combat experience, and a stint at Edwards as an experimental flight test pilot—redefined the status hierarchy of the astronaut corps. No longer were all astronauts created equal, and only those with backgrounds similar to Slayton's were considered for the most prestigious flying opportunities.

Slayton soon received the managerial assistance of a bona fide space hero, Alan Shepard, whose bid for a long-duration mission was overtaken by the sudden onset of an inner ear disorder—Ménière's disease—that grounded him in 1963. Shepard's misfortune provided Slayton with a "lieutenant": a "keeper of the traditions and rules" of the Astronaut Office, who junior astronauts found to be a capricious and often terrifying presence at NASA. With Slayton as director of Flight Crew Operations and Shepard as chief of the Astronaut Office, the two men together corralled, guided, and, later, protected their colleagues from embarrassment.[183]

Having established the ability of humans to survive in space, NASA expanded, invigorated by an administration enthused by Shepard's flight and determined to answer Soviet achievements with a high-profile project that would assuage domestic critics. As early as 1959, a NASA planning committee had chosen lunar landing as the nation's long-term spaceflight goal, though funding to support it was not forthcoming from the Eisenhower administration and NASA placed the goal of a permanent orbital station ahead of piloted landing, relegating the latter to the 1970s.[184] To Wernher von Braun, who Vice President Lyndon Johnson had solicited for his advice on spaceflight matters, piloted lunar flight represented such a difficult technical undertaking that the United States had a "sporting chance" of beating the Soviets if it chose to pursue this goal.[185] NASA, after 1961, set quickly to work building the hardware needed for the voyage and recruiting the pilots needed to man it: 30 by the end of 1963.[186] The Original Seven cast a long shadow over "Deke's boys," selecting, training, and assigning them to flights, and seeking to perpetuate the high standards they had set—and often could not maintain—for themselves.[187]

Chapter 2

"Deke's Boys"

The success of Project Mercury emboldened President Kennedy to challenge the National Aeronautics and Space Administration (NASA) to pursue a more audacious goal than Earth orbital flight; this augmented effort required not only new hardware and facilities, but a substantial increase in the size of the astronaut corps, only one year into NASA's human spaceflight program. While the culture of the astronaut corps emerged relatively quickly after the selection of the Project Mercury astronauts, the Astronaut Office's establishment as a true professional institution occurred only with the arrival of more junior spacefarers in 1962 and 1963. Upon the arrival of these (and subsequent) rookies, veterans found themselves charged with not only their own training, but with training and supervision of new colleagues. Now able to reproduce themselves, astronauts became a professional community with an explicit hierarchy, standards of conduct, rites of passage, and an operational lore that could be passed down from generation to generation.

Astronauts joined NASA in the 1960s, aware that within a few short years they might very well find themselves on the Moon, or dead, or possibly both. The potential for glory and the likelihood of violent death were only two of the factors conditioning the astronauts' expectations and constraining their behavior. Indeed, multiple overlapping, interconnected regimes of control disciplined the astronauts' working lives, some of which they could influence and others that they could not. If comparable to engineers or industrial laborers in their discontents, astronauts were unlike other workers of their era in their celebrity, which opened them and their working environment to constant scrutiny. Together, these forces produced, for the astronauts, a work culture often at odds with their public appearance.

Expansion of the Corps

Prior even to the conclusion of Project Mercury, NASA recognized that a schedule of spaceflights leading to a piloted Moon landing required far more flight personnel than it possessed. By 1963, four of NASA's seven astronauts seemed likely to never fly again; in order to land on the Moon, though,

NASA contemplated dozens of multiperson flights to perfect procedures and technologies for deep space navigation. Project Gemini contemplated ten missions using an enlarged Mercury capsule augmented with a second crew position and more robust systems for orbital maneuver and long duration operation. Project Apollo, to follow Gemini, would require three-man crews, greatly increasing the number of astronauts required for prime and backup crew positions.

To fill the positions of Gemini and Apollo crews, NASA selected, during the 1960s, new groups of pilot-astronauts roughly every other year, typically choosing from several hundred or more highly qualified applicants for a dozen or so spots (Figure 2.1). Like the astronauts of 1959, applicants in later selections endured a battery of physical and psychological tests; in these evaluations, though, veteran astronauts hovered over the process, particularly Deke Slayton, who interviewed applicants along with other members of the Selection Board. The evaluations mixed objective criteria with subjective impressions; even as late as 1966, performance in the interview stage constituted 18 of the 30 potential points awarded to applicants, swamping academic qualifications and even flying performance as evaluative tools.[1]

Throughout the 1960s, the astronauts remained a largely homogenous group of military pilots. The 1962 selection consisted entirely of professional

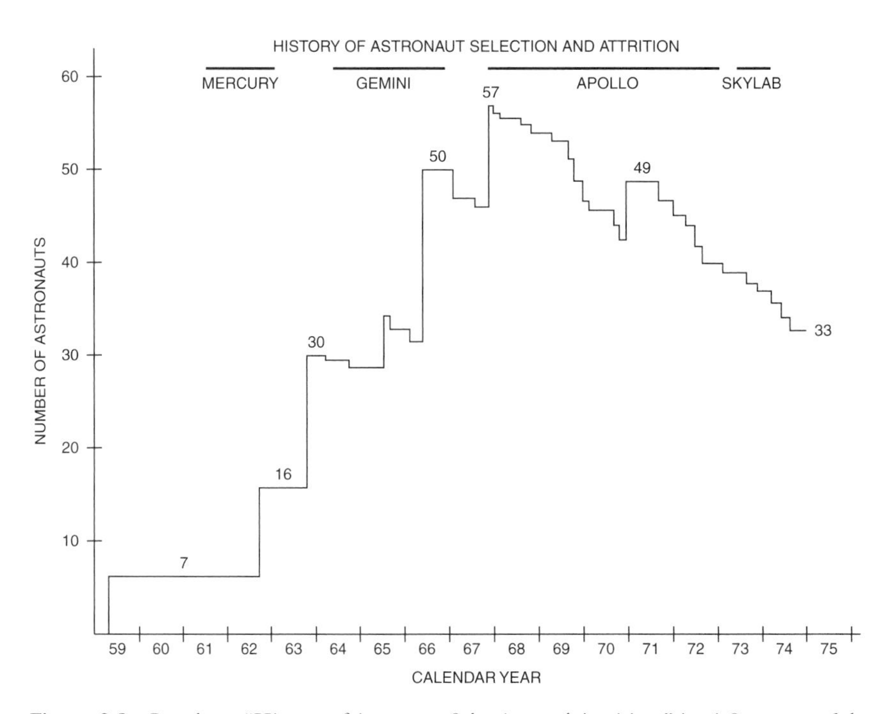

Figure 2.1 Based on, "History of Astronaut Selection and Attrition," in *A Summary of the Astronaut Recruiting and Selection Process* (Houston: NASA, 1975), 18 (Box 10, Charlesworth Files, Center Series, Johnson Space Center History Collection, University of Houston–Clear Lake archives)

test pilots, while operational pilots joined the ranks beginning in 1963. The test pilot community was so small that many applicants were already personally known to members of the astronaut corps at the time of selection. Astronauts of the 1959, 1962, and 1963 selections especially were close enough in age to be contemporaries; many participating in the 1962 and 1963 selections had also applied in 1959 or 1962, or had served with those selected at either of the two military test flight schools. At Edwards Air Force Base, for example, Michael Collins (1963) had trained with Frank Borman (1962), Charlie Basset (1963), and Ed Givens (1966), and received instruction from Tom Stafford (1962), who encouraged Donn Eisele (1963) to join the Astronaut Corps.[2] Wally Schirra (1959), Pete Conrad (1962), and Jim Lovell (1962) studied together at the navy's Test Pilot School; Alan Shepard (1959) had instructed Pete Conrad (1962), who had shared a berth on the USS *Ranger* with Dick Gordon (1963) and later instructed Alan Bean (1963).[3]

If the Original Seven had been junior military test pilots unprepared for the celebrity and workplace culture they encountered, NASA astronauts of 1962 and 1963 were men well suited to administration and public relations. Describing the "average astronaut" in a 1965 press release, NASA seized upon his professional credentials: the typical astronaut, NASA reported, "holds a bachelor of science degree and a master's degree, and has done some work on his doctorate. He graduated in the top five from a major American university, and attended military test pilot school." These men were experienced aviators with thousands of hours in the cockpit but, the release noted, there were no combat "aces" among them.[4] Instead, they had distinguished themselves through their leadership potential and social acumen. Conrad was a Princeton University graduate; the air force's Tom Stafford joined NASA three days after matriculating at Harvard Business School to pursue a master's degree in business administration; Frank Borman was an air force test pilot instructor with hopes of a government career; and Jim Lovell was an affable naval aviator with a life-long interest in rocketry, who Collins described as better suited to the "PR world" than to engineering. Even the two "civilian" pilots in the 1962 selection (Neil Armstrong and Elliot See) had enjoyed distinguished military careers. The astronaut class of 1962, in particular, proved NASA's most accomplished, counting among its members virtually all of the men who commanded the Gemini and Apollo flights of the 1960s and 1970s.[5]

For these men and for those who selected them, space work presented fewer unknowns than it had for the Original Seven: unlike the Mercury "volunteers," these flyers had the benefit of joining a program with an established goal, a brief but encouraging record of success, and no fatalities to date.[6] They feigned no confusion, displeasure, or ambivalence about the space program and pursued it with naked ambition: Collins, a young air force test pilot in the Fighter Branch at Edwards, had his "application in before the ink was dry." Compared to hurtling through space, terrestrial test pilot work, which only a few years earlier had seemed to Collins the most satisfying and rewarding job in the world, was no longer "larger-than-life." When he failed to make the cut for the 1962 selection, and despite receiving no encouragement to do so, Collins applied again, successfully, in 1963. Seeking to ensure that air force flyers dominated the Astronaut Corps, the service, in 1961, redesignated its test pilot training program as the Aerospace Research Pilot School, specializing in training pilots in spaceflight operations and astrodynamics. (The air force also instituted a "charm school" to drill pilots in the social graces they needed to survive the NASA interview process.)[7] Early graduates of the School, like MIT-trained engineer and test pilot David Scott, were significant members of later selection groups.

On the Job

With the astronaut corps growing so rapidly in the early 1960s, even Walter Cunningham, an astronaut of the 1963 selection, could experience the giddy sensation of arriving at NASA as young neophyte and, only five years later, wander around NASA's hallways as a veteran space pilot able to haze and mentor new arrivals. Literally flying into Houston in an A-4D Skyhawk borrowed from the Marine Corps Reserve, Cunningham, on his first day at NASA, had parked his jet next to the NASA planes and strolled into the flight office of Ellington Field, admiring the astronauts' custom-made blue coveralls and waving presumptuously to the already-legendary John Glenn and Alan Shepard. At least initially, Cunningham's delight at "joining the last of the world's great flying clubs" outweighed his feelings of inferiority at being a relatively late arrival to the astronaut corps. Throughout the 1960s, junior spacemen like Cunningham could, after a year or two as trainee, look to the arrival of a "new" bunch of rookie astronauts whose appearance made the pledges finally feel "real."[8]

From the runways at Ellington Field, astronauts soon found themselves at Houston's Manned Spacecraft Center, a stretch of nondescript buildings constructed on what Norman Mailer described in his 1970 book *Of a Fire on the Moon* as "flatlands behind a fence," 25 miles south of Houston. To Mailer, the "all-but-treeless" site was instantly recognizable: a "geometrically ordered arrangement" of modern buildings, "milk-of-magnesia white," "severe, ascetic, without ornament, nearly all of two or three stories"—in short, "buildings and laboratories which seemed to house computers, and did!" The compound could pass for "an industrial complex in which computers and electronic equipment were fashioned," or a "marvelously up-to-date minimum security prison"—mostly, though, it conjured up a disquieting future: "the worst of future college campuses," Mailer writes, "miserable," "brand new," and with "a general air of business administration."[9] For astronauts, though, it was an improvement over the ramshackle quarters of Edwards Air Force Base, and for them it would be home for their entire tenure at NASA.

While the job of astronaut required a variety of administrative and public relations duties common to other professional environments, it was, in a few critical ways, very different, and astronauts both recognized and exploited the differences to improve their station. Few engineers outside military service expected to die violently on the job, and the astronauts' awareness and seeming comfort with their own mortality marked them as a special breed of human, warranting a certain indulgence with regard to their personal behavior. Nor were many other American engineers of the period expected to serve as propaganda tools, their identities exploited by the government to promote American power oversees. Astronauts of the early 1960s, though, quickly found themselves juggling competing responsibilities and identities, aware that any autonomy they enjoyed was dependent upon both their success in space and their behavior on the ground. Whether in orbit or on Earth, their days were choreographed sets of activities, observed by nearly everyone above and below them in rank, from the president of the United States to school children on distant continents. On the ground—where astronauts spent virtually all of their professional lives—offices, classrooms, factories, cockpits, and countless public relations visits offered a work environment as hazardous—literally, considering their constant jet training—as the one they encountered in space.

Once selected, astronauts were quickly absorbed into NASA's training, technical, public relations, and administrative operations at MSC, work that consumed virtually the entirety of their working life. Flying in space for these men was practically an incidental activity, taking up so little time it hardly registered on their schedule. Spaceflights through the 1960s lasted, on average, only a few days, and astronauts often waited several years between flights. NASA's two most influential astronauts of the 1960s, Shepard and Slayton, had 15 minutes of spaceflight between them. If astronauts seldom flew, though, they were nearly always thinking about flying. Everything, from design work to commuting in airplanes they flew themselves was, somehow, preparation for the event that would be the focus of their careers.

Although all or most had joined the program in the hopes of actually flying, classroom work, simulator time, proficiency flying in jet trainers, survival training, and engineering assignments occupied the vast bulk of their time for the first two years as astronauts (Figure 2.2). Much of the astronauts' training adopted quasi-military character, as pilots maintained jet proficiency, kept physically fit, and prepared for unlikely emergencies, including wilderness survival in every kind of ecosystem they intended to overfly (Figure 2.3). A pilot's initial months in the Astronaut Office typically entailed an elaborate series of lectures and training courses in spaceflight theory and practice, vehicle systems, and survival in all of the various climes astronauts and their spacecraft might someday find themselves if they de-orbited at the wrong time. At times, astronaut and astrodynamics expert Buzz Aldrin was galled by his colleagues' intellectual stupor; astronauts insisted that classroom instruction never be graded, so the pilots would not have to worry about yet another matrix on which they might be compared.[10] Graduation from this spaceman boot camp meant time for proficiency flying, assignment to engineering work and, for a fortunate few, crew training.

For the astronauts, though, their favorite training, and the one that accorded them the most status, was routine flights aboard the T-38, a two-seat

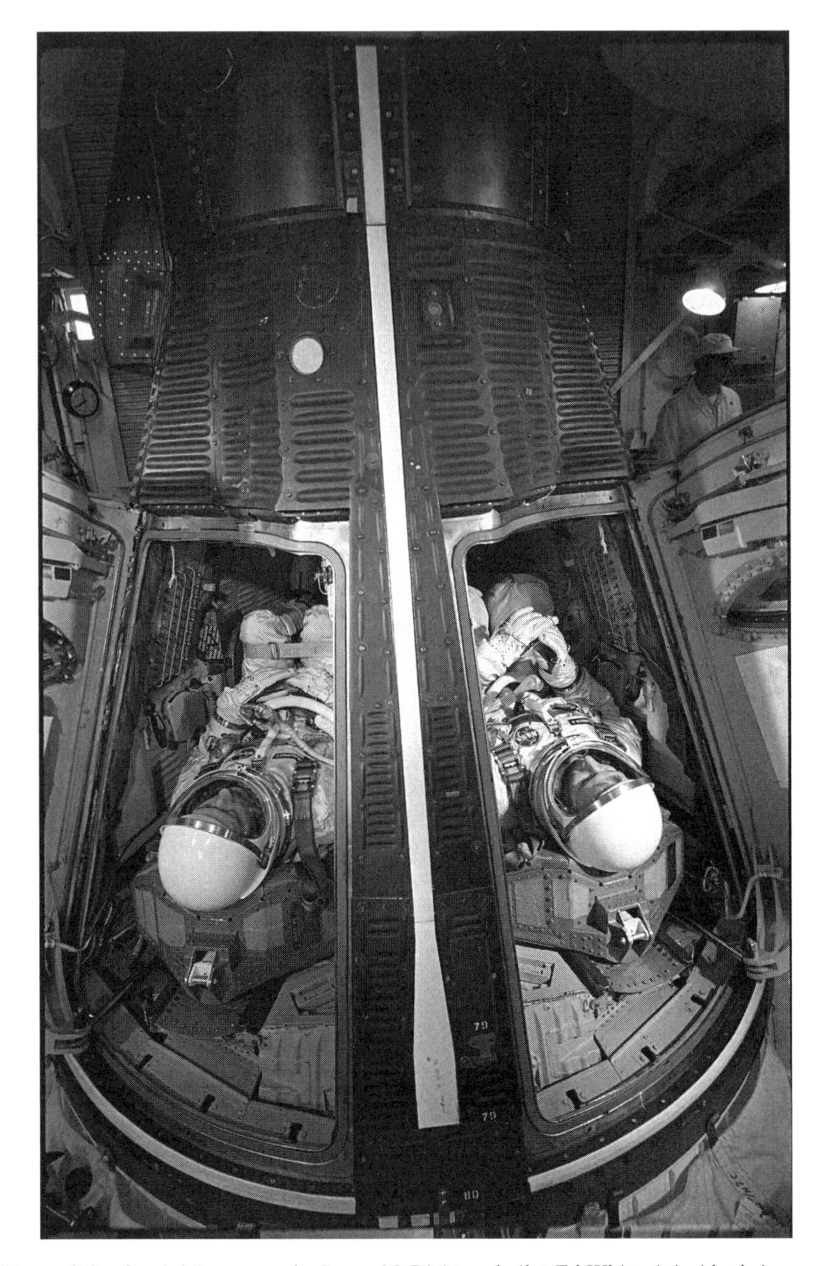

Figure 2.2 Gemini 4 commander James McDivitt and pilot Ed White sit inside their spacecraft for a simulated launch at Cape Canaveral, Florida in 1965 (NASA photo)

jet trainer capable of flying at twice the speed of sound and a recent introduction to the air force inventory in the mid-1960s. Astronauts, to maintain their proficiency and accomplish their duties, eventually enjoyed unrestricted access to a fleet of T-38s, a major perk that set them apart from other government pilots. The astronauts loved to fly the nimble, slightly cantankerous

Figure 2.3 Gemini 5 Commander L. Gordon Cooper Jr. (in raft) and pilot Charles "Pete" Conrad Jr. (in water) practice survival techniques in the Gulf of Mexico prior to their 1965 flight (NASA photo)

jet, and did so constantly. For many, their schedules required almost daily travel, always by a jet they flew themselves from their homes in Houston at the Manned Space Flight Center to Los Angeles, St. Louis, or Washington DC. In the space of a few short years, though, four astronauts were killed while flying T-38s, due to a bird impact, delayed ejection, cloud cover, and an unexplained loss of control. For jet pilots, fatalities in training were unremarkable, if tragic events. The astronauts expected that some of them would die prematurely, and seemed little surprised when they did. Charlie Bassett, who died in landing crash in a T-38 flown by Elliot See, "was a remarkable man," Collins writes, "and then he was decapitated in a parking lot." For Collins, though, a plane that "took one good man from us every two years" still "gave more than it got."[11]

If living the dangerous life of professional pilots, astronauts insisted that they represented critical members of the space program's engineering community. Upon joining, astronauts received technical assignments—communication, spacesuits, orbital rendezvous—and worked with NASA scientists and engineers on these projects while preparing for flights. These efforts, as well as the innumerable public relations duties, constituted much of the work of astronauts when they were on the ground. For certain astronauts, the ideal engineer-astronaut was both a team member and an aggressive leader, devoted to mission success and unwilling to accept the assurances of others. Astronaut Roger Chaffee, Grissom once recounted, was one of the best: "When he starts talking to engineers about their systems, he can just tear those damn guys apart. I've never seen one like him. He's a really great boy." This aggressive attitude, though, occasionally earned the pilots the ire of ground engineers, and astronauts struggled to balance professional interests with their duties to the NASA hierarchy. Apollo 7 commander Wally Schirra and his Command Module Pilot (CMP) proved particularly aggressive. "[NASA] regarded us as troublemakers," crewmate Donn Eisele noted in a later interview.[12]

The astronauts' engineering work legitimated their high-profile presence in the human spaceflight program. New astronauts arrived in a workplace in which all pilots, no matter how junior, were expected to assume leadership responsibilities over technical teams to which they were assigned, even if they knew almost nothing about the area of research. The rare junior astronaut, Alan Bean recalled, knew his specialty well. Fellow 1963 selections Aldrin and Scott were technical experts, and 1966 astronauts like Charlie Duke and Ed Mitchell were deceptively sophisticated on technical matters. When visiting MIT in 1962 for a briefing on Apollo's sophisticated guidance computer, Armstrong asked "smart questions" demonstrating his vast experience with computer control in the X-15 program. For such men, NASA provided the opportunity to play a meaningful role in refining designs.[13]

Most other astronauts, Bean figured, probably just faked their way through endless meetings, mocking radical ideas they didn't fully understand. Bean, uncomfortable with such self-delusion, struggled when assigned to assist in recovery planning. "I knew I was a good pilot and I knew how to do Navy things, but when I got in these meetings at NASA with people that had been in recovery for ten years, I didn't really feel like my ideas had a lot of merit, because I really didn't know anything about it. I had had no recovery training, and I'd go to the meetings, I really didn't know what to do." "So I would go into a meeting, like in this room, and people would look at me like, 'We're waiting for you to tell us something.' Well, what the hell? . . . So I'm sure it was a big disappointment to the group because I didn't have anything to tell them. If I did have something to tell them, I was afraid to do it."[14]

Astronauts, in addition to contributing to engineering teams, were expected to serve as spokesmen for the very engineers they privately criticized, touring plants to motivate workers and speaking at public events to encourage support for the space program.[15] (Bassett and See, notably, died in plane crashes while commuting to a McDonnell Aircraft plant in St. Louis.) Astronauts toured the country incessantly at the request of the press, civic organizations, schools, and other institutions, yet the process by which NASA dispatched the men was semirandom. Entities solicited the appearance of well-known individuals in writing to NASA administrators, but known and unknown astronauts alike alternated speaking duties with a weekly rotation. On duty for a given period, an astronaut might deliver "the same speech thirty times," in different locations, chaperoned by a NASA Public Affairs handler to keep him on schedule.[16] For astronauts, public relations work extended to astronauts' families, with one astronaut routinely assigned as a "family escort" to interact with the astronauts' wives when mission duties occupied their husbands.[17]

While certain astronauts were known for their social ease, others regarded the endless public relations work as an annoyance. When asked to communicate more than the mechanics of their profession, many struggled. Some, like Slayton, were hopelessly profane; others, shy and awkward.[18] NASA saw to it that astronauts received ample briefings on NASA's goals and intentions, but while all of the astronauts could be glib for short periods of time, protracted press interviews and tours taxed certain of the astronauts' rhetorical skills. NASA managers, though, provided relatively little instruction on content and tone, and preferred to see the astronauts script their own speeches; for Christmas eve 1968 radio broadcast from the historic Apollo 8 circumnavigation of the Moon, NASA Office of Public Affairs chief Julian Scheer's instructions to Commander Frank Borman had been only to say something "appropriate."[19] Reviewing this policy in November 1972, NASA assistant administrator for Public Affairs John Donnelly reiterated that the astronauts' comments should be on message, but original. "They may not be as poetic in all cases, but I think there is no substitute for sincerity and realism as expressed in a man's own words," a conclusion to which deputy administrator George Low responded with one handwritten word: "agree."[20]

Combined, the astronauts' training, engineering, and public relations duties consumed virtually all of their time, at levels of compensation that were never more than adequate. With its constant travel and frequent fatalities, the astronaut lifestyle could have proven extremely challenging for young family men if not for the indirect compensation they received by trading on their fame. As astronauts might spend virtually their entire workweek away from home, they struggled to subsist on the $16 per day allotted to them by NASA for expenses, a ceiling astronauts stretched by gratefully receiving unpublished "astronaut rates" on accommodations. Fueled by perks, astronauts enjoyed a "standard of living and social life that greatly exceeded" their income, but without the publicity contracts, outside directorships, stock options, and real estate deals many astronauts cultivated, a rookie astronaut's salary of $13,000 in 1963 (approximately $90,000 in 2010) was not enough to support housing, family, and travel expenses.[21]

Though astronauts remained open to press interviews regarding their professional activities, the earliest astronauts doubled their salaries by negotiating deals that gave certain publishers exclusive access to their "private stories," which included their family histories and subjective impressions of spaceflight.[22] The commercial relationship established between the astronauts and the publishers was the source of controversy almost immediately; upon the astronauts' arrival in 1959, NASA Office of Public Information chief Walter Bonney had, in consultation with the Department of Defense (DoD), determined to allow the astronauts to sell their stories with NASA's approval, ensuring that the Agency retained control of the astronauts' public image.[23] The idea that the astronauts would negotiate individually with publishers was an anathema to Bonney, who arranged for Mercury astronauts to retain entertainment attorney Leo DeOrsey to negotiate for a group publicity contract for the astronauts and their spouses, eventually executing an agreement with Time, Inc. that resulted in numerous stories in *Life* magazine.[24]

Upon taking office in 1961, President Kennedy was at first firmly opposed to the relationship, as was much of his staff (including Pierre Salinger, McGeorge Bundy, and Theodore Sorenson), many in the DoD (including Cyrus Vance, the DoD's general counsel), and James Webb, the new NASA administrator.[25] In one conversation with President Kennedy, though, John Glenn, lobbying for his peers, likened astronauts to soldiers who must accept, in addition to the dangers of public service, the indignities of public notoriety.[26] Kennedy was apparently swayed by the argument, but more important to the deliberations over renewal of the *Life* contract was NASA's realization that it had yielded abundant, positive press coverage about the astronauts and had made the men and their wives "manageable," preventing them from competing for magazine deals or publishing negative, conflicting, or contradictory information.[27]

Rather than bar the contracts, NASA eventually restricted their application to the "family stories" of the astronauts.[28] By 1973, even Webb had turned from a critic of the contracts to an advocate, writing (in response to an exposé on the publicity deals by Robert Sherrod),[29] that the contracts had protected the astronauts' families from excessive press scrutiny and had eliminated potential competition between the men for publicity, more of NASA's concern than that of the astronauts.[30] With the arrival of new astronauts in the early 1960s, publicity contracts soon multiplied between the astronauts, *Life*, and Field Enterprises, publisher of the *World Book Encyclopedia*. The astronauts also received life insurance policies worth $100,000 from Field.[31] Most of the astronauts had been unable to secure life insurance after joining NASA; NASA declined to cover astronauts under its group plan due to the extreme cost involved, though active duty military pilots could still claim coverage under their service policies.[32]

The publicity contracts added another $16,250 to Cunningham's initial annual salary, a bonus for which he and the other astronauts were grateful. The contracts, though, served NASA's interests even more than those of the astronauts: it bolstered astronaut salaries at private expense, limited negative publicity, and moderated the behavior of astronauts and their spouses, all of whom were obligated to release personal information through *Life*, or not all. Even Life and Field were sometimes distressed by arrangements that, ultimately, served NASA's interests above all others. With interviews vetted by NASA, publishers often complained about a lack of access and candor in articles, yielding dull expositions in which astronauts invariably appeared "deodorized, plasticized, and homogenized."[33] A book of vignettes by the Mercury astronauts published by Simon & Schuster (the rights for which it purchased from *Life* for $200,000), entitled *We Seven*, proved equally unsuccessful. Nevertheless, the rights to the astronauts' stories remained valuable commodities through the end of the 1960s.[34]

The astronauts, though, freely competed for other financial benefits of their office. NASA prohibited the astronauts from seeking outside consulting or speaking work (or work for government contractors) but the agency was unable to police the use of corporate jets, party invitations, invitations to sit on corporate boards, or insider stock market tips that seemed to cross the astronauts' desks everyday, often from unscrupulous businessmen. What made such deals possible was the astronauts' celebrity, valuable as a marketing or negotiating tool, and astronauts became comfortable ingratiating themselves with wealthy men eager to rub shoulders with spacemen.[35]

Below the big dealings were many more gifts and sweetheart deals for mortgages, cars, and other extras, some of which NASA was able to extinguish and others which it could not. (Shepard, in particular, confessed to one NASA investigator about receiving a plethora of silver bowls, golf clubs, and other gifts.[36]) Astronauts particularly adept at "scamming" (as Schirra is reputed to have called it) found themselves habitually receiving all manner of benefits.[37] NASA prevented some of these arrangements but enforcement of the rules often fell to Shepard, whose own prowess at cultivating business contacts was particularly respected by the other astronauts. When challenged about his own dealings, according to astronaut Joe Allen, Shepard yielded no ground, suggesting that complainers "talk to my friend the President" if they had any questions.[38] Corvette sports cars provided to astronauts at dealer cost were among the most prized benefits. An investigation of the astronauts conducted in August 1972 revealed a longstanding relationship between Florida and Texas automobile dealers and astronauts. Astronauts, one NASA investigator determined, assumed that Shepard—NASA's most senior astronaut and the *de facto* head of the Astronaut Office—represented "NASA Management," and that, because he had enjoyed this perk, others could, as well.[39]

Car dealers, politicians, businessmen, and socialites, Cunningham recounted, eagerly "collected" astronauts, and many were eager to receive their hospitality.[40] Unsophisticated in the commercial world, though, many astronauts were swindled or had their reputations tarnished by association with shady business dealings. The simplest of these involved the sale of flown objects to collectors, who eagerly snapped up virtually anything the astronauts brought back from space. NASA eventually permitted the astronauts to "carry personal souvenir-type items" aboard the vessel, on the condition that the astronauts disclose their presence and not use them for "commercial or fund-raising purposes," but the agency had difficulty policing the rule.[41] Some more ambitious astronauts (Shepard, in particular) avoided the disrepute associated with souvenir sales in favor of business partnerships and real estate deals and, Cunningham claimed, became millionaires on NASA's time.[42] Rarely, though, did these dealings cross into the kind of "influence peddling" that stood to jeopardize the program; less constrained, though, were the personal relationships astronauts often undertook when socializing after work.

5 O'clock Shadow

To some astronauts, space piloting was a profession demanding faithfulness to their public image virtually round-the-clock. "Even early in the project," Korchin and Ruff noted, "some of the men were quicker than others to realize that the astronaut's role carried a social responsibility and to feel obliged to live up to the ideals of such a role. Like Caesar's wife, they felt they had to be above reproach."[43] Others, though, saw themselves as artisans who could not be expected to temper their recreational behavior to suit their employer's ideas of morality. Like the craft workers who sought to maintain their often raucous drinking culture in the face of new forms of factory discipline,[44] many astronauts freely partook in extramarital affairs and other cavorting, a split that astronauts were unable to conceal even from the psychiatrists. "For others of them," the psychiatrists wrote, "the astronaut's role was more narrowly defined, anchored in the piloting job. These men saw themselves as craftsmen who had to do an important job as well as possible, but what they were or did beyond that was wholly their concern."[45]

By 1963, veteran astronauts just assumed that new hires would take advantage of their positions to enjoy sexual favors from celebrities, random acquaintances, and the space groupies who hovered around Cocoa Beach, Florida—a weird predatory culture that produced some odd encounters and surreal locker-room talk. Slayton warned newly selected astronaut Cunningham and his colleagues that they were "big boys now." They would "all...get a lot more play from the girls" and if they planned on "screwing around" they would "better be damn discreet about it." True to Deke's warning, the infidelity of many astronauts was an open secret within NASA, but only public embarrassments triggered official condemnation. If the astronauts stayed out of the newspapers, they could enjoy the benefits of their notoriety. "[A]n astronaut generally expects a woman to accept if he asks her out," Cunningham recalled, and "she generally does." A private nighttime cockpit tour of the astronaut's T-38 jet—the "golden leg spreader"—often ensured a coupling.[46]

Alcohol, traditionally imbibed to excess by pilots, often fueled the astronauts' escapades. Wolfe, in *Right Stuff*, described early astronauts as a hard-drinking lot, one aspect of test pilot culture the astronauts brought with them to NASA. Off-duty, military pilots in remote bases drank whatever inexpensive alcohol was available nearby and then tested their mettle in automobiles. Often, these escapades lasted all night, with the pilots, still intoxicated, quietly dragging themselves to their planes the next morning, hoping that the oxygen flowing through their masks would "burn the alcohol out of their systems." Once ensconced in Florida or Houston, astronauts easily replicated this lifestyle, supplemented by endless party invitations. NASA, while wary of the astronauts' celebrity, could exploit the concern over their public reputations to moderate the pilots' rambunctious extracurricular behavior. Rather than suppressing their "flying & drinking" culture outright, NASA acknowledged and attempted to moderate it, encouraging astronauts to enjoy a hard drink at social occasions, but enjoy only one.[47] Outside the astronaut's judgment and the firm and omnipresent hand of Robert Gilruth, though, few acted to restrain the fraternity culture that astronauts of the 1960s brought with them, especially local law enforcement, which turned a blind eye to the speeding and drinking present wherever astronauts assembled. The press chose to suppress most such reports.[48]

Space work, often accompanied by fear of death, frequent absences, and drunken bouts of serial adultery, was hard on marriages. Cunningham recounted being absent 265 days in a single year, straining and distorting his home life.[49] By his account, 18 percent of astronaut marriages ended in divorce within 15 years, including six divorces within one two-year period around the time of the first Moon landing, "leading some critics to suspect that we were coming back from space with some kind of social disease or a post orbital melancholy."[50] (Indeed, Frank Borman, in 2008, recounted with pride at commanding the only Apollo crew still married to their first wives.[51]) The pressure of spaceflight was not so much in causing divorce as in delaying it, forcing astronauts and their wives to cling to unhappy unions for the sake of appearances.[52] NASA had considered interviewing applicants' spouses during the first astronaut selection but had decided against it, assuming the pilots would self-report accurately on their personal relationships and that other screening tools would weed out those with unsupportive home lives.[53] Neither assumption was true.

While their husbands worked, traveled, and caroused, the astronauts' wives cobbled together their own social networks to support themselves and their children. NASA encouraged these support networks; marital instability threatened the reputation of the program, and astronauts whose domestic difficulties threatened to spill into the public eye were encouraged to temper their behavior or resign.[54] Prior to 1965, astronauts had enjoyed the protection of NASA officials, and reporters tended to minimize any stories of domestic problems or reckless personal acts. This protection, though, had limits—in 1965, promising astronaut-candidate Duane Graveline was dismissed from NASA only weeks after selection, when an acrimonious divorce confronted NASA with what some feared would be ghastly publicity.[55] The firing chilled other astronauts whose marital difficulties and flamboyant social activities had already become legendary within the small community.[56] While NASA's enforcement of the no-divorce rule eventually weakened, marital problems that affected an astronaut's job performance were still serious matters.

If the celebrity of the astronauts extended well beyond the workplace, then so did the systems of control that modified astronaut behavior and established limits for what could and could not be done. In 1968, Apollo 7 astronaut Donn Eisele, struggling with a failing marriage and an extramarital affair, lost the confidence of his superiors due to what they perceived as lackluster enthusiasm and erratic attention to detail. Eisele's marriage had been a parade of suffering and alienation; frequently absent, he struggled with his son's leukemia and death at the age of six. At some point prior to this, he had begun seeing a woman in Florida while concealing his marital status, which infuriated his paramour and soon came to the attention of his wife. Deke Slayton, concerned with Eisele's upcoming flight, encouraged

Eisele to maintain both these relationships, going so far as to call Eisele's mistress, Susie Hearn, to encourage her to "hang in there until after Apollo 7 flew." During the mission, a frazzled, sleep-deprived Eisele had berated ground controllers and argued openly with Schirra. Apollo 7 doomed all three crew members to abbreviated flying careers, even with Stafford's strong support for Eisele. (Finding that Eisele, upon his return, had been assigned to join Gordon Cooper's Apollo 10 backup crew, Cooper was crestfallen: sharing a slot with Eisele all but ensured his crew would never fly.) Eisele was not the first astronaut to begin a clandestine affair, but in divorcing his wife on his return to Earth (and marrying his Florida girlfriend), he violated an unwritten rule that the astronauts' Florida social lives would never overlap with their home lives in Houston. Eisele's new wife was shunned by the astronauts' spouses, and he soon left NASA, to the relief of his adulterous colleagues.[57]

Though partaking in the recreational and social opportunities their celebrity afforded them, at least a few astronauts seemed conflicted by the degree to which their upstanding public images deviated from their imperfect and often decadent private lives; deep down, Cunningham wrote, even the wildest astronauts feared they were, "at the bottom...dull men, sent to do a technicians job," whose temporary astronaut status provided them with a "life-style that in other places or other times would never have been accessible."[58] That many astronauts declined participation in boozing and adultery only made the revelers more uncomfortable, piercing the rationalization that such behavior was normal or justified given the stress they were under. While bachelor astronauts like Jack Swigert were never seen "with the same woman twice," Borman, Ed White, and others were so faithful and sober that the "swingers" resented them; and encouraged the devoted family men, unsuccessfully, to stray. Though astronauts shied away from overt moralizing, the "straight arrows" made the "wilder ones" feel guilty merely through their forbearance, especially when one faithful astronaut refused to indulge in a "particularly tempting dish" his colleagues had ordered to his hotel room, in a half-baked attempt at seduction.[59]

Race to Space

Veteran space pilots, Cunningham noted, attracted the best groupies: "specialists" who spurned the "unflown rabble" for newly returned crews.[60] For most astronauts, the honeymoon they enjoyed upon joining NASA quickly subsided, giving way to an awareness that a sharp distinction existed between those who had flown in space and those who had not, a principle Cunningham called "John Young's Law," after the 1962 astronaut who was first among his peers to fly. New astronauts' rookie status invaded every aspect of their professional life, never more so than during public relations work, where veteran crews received the best invitations and most attention. Expecting heroes, Cunningham recounted, audiences sighed in disappointment when the astronaut NASA provided was a rookie with no thrilling space stories to tell, forced to share motion picture footage of a colleague's recent flight.[61] Fresh from a spaceflight, though, astronauts could enjoy immediate notoriety, validation, the opportunity for advancement. Flight brought more income and new responsibilities; astronauts on active duty in the military usually received promotions upon their return, ensuring them higher pay and better benefits from their respective services.[62]

Upon their return from their first mission in 1965, Jim McDivitt and Ed White received almost instant promotions from President Johnson. Their promotion created a moment of awkwardness for other, veteran astronauts, which Johnson ultimately resolved by fast-tracking the promotions of earlier crews, a stopgap measure that suggested the need for a more formal policy. While the navy and air force preferred "that the promotion of astronauts take place in accordance with the normal procedures of the respective services," Johnson demanded a "consistent policy."[63] By the end of the year, NASA and the DoD had worked out an arrangement, formalized by Executive Order, that provided military astronauts with a decoration and promotion upon completion of their first successful spaceflight, through the rank of Colonel (air force, marines) or Captain (navy).[64] (NASA eventually implemented a similar policy for civilian astronauts, providing promotion through GS-15 for spacecraft commanders.[65]) A subsequent directive from President Nixon amended the policy to provide astronauts with a second promotion upon completion of a lunar or interplanetary flight, a policy that was extended by President Reagan and remained in effect through 1985.[66] When solicited for its opinion on matters of promotions outside the normal guidelines, NASA's official policy was to defer to military authorities on the subject.[67]

Astronauts, more than anything else, wanted to fly, and anxiously waited their turn for the crew assignments that would catapult their careers, and without which their time at NASA could quickly become unbearable. (The concern was particularly acute among the astronauts recruited after 1962, for whom the possibility that they might never fly constantly hovered in the background.[68]) By 1963, most of the 1959 group astronauts had flown in single-seat Project Mercury vehicles; between 1965 and 1966, members of the 1959, 1962, and 1963 groups flew in the two-seat Project Gemini spacecraft, providing them with training and experience that placed them at the "top of the pyramid" for flying opportunities in the Apollo program, set to begin in 1967.

The lore of astronaut assignment process during the 1960s at NASA remains shrouded in controversy despite its almost complete lack of real mystery.[69] In Project Mercury, astronauts, like other workers, had labored under the "simple" discipline of NASA managers to whom they reported and who possessed the power to terminate their flying careers almost instantly.[70] Astronauts could hope to influence some of these individuals, but their powers of persuasion were limited and could be brought to bear only infrequently. Though authority for crew assignments, throughout the 1960s, rested principally with Robert Gilruth, Slayton enjoyed principal responsibility for recommending astronauts to crew positions beginning with Gemini.[71] NASA

Headquarters, according to Collins, "always rubber-stamped Houston's recommendations."[72] Indeed, NASA, according to Apollo program manager George Low, was seldom able to change Slayton's mind when he proposed a crew;[73] "[w]ith few exceptions," wrote Cernan, "Deke always made the final call on who flew and in which seat."[74]

Slayton claimed to have arrived at assignments informally ("over a few beers" with Shepard), that no grand systematic scheme determined who would fly or when.[75] Even Gilruth later claimed to not to know how Slayton arrived at his crew rosters.[76] Junior astronauts, according to Cernan, "studied his choices with great care, looking for some pattern," but eventually convinced themselves that "there wasn't one."[77] Slayton indicated in his memoirs, though, that crew assignments were the subject of a very simple calculus, and when Gilruth questioned Slayton on his choices, Slayton always responded satisfactorily.[78] While Slayton repeatedly insisted that all astronauts were qualified ("if I hired a guy and kept him, he was eligible to fly"), he freely admitted in his 1994 memoir that "some were more qualified than others."[79] Describing his assignment "guidelines," Slayton conceded that he had reserved "challenging missions" for astronauts with "command or management or test pilot experience," attempting to build an experienced workforce for future programs. Slayton, Neil Armstrong later speculated, chose crews chiefly to cultivate future commanders, subordinating other considerations to that goal.[80] Indeed, Slayton noted this in his memoir, writing that he "always kept future requirements and training in mind."[81]

The growing number of astronauts and finite number of flights, revealed the essential dilemma of crew assignment—the presence of experienced astronauts offered missions a greater chance of success, but failure to fly newer astronauts undermined morale and decreased the pool of experienced astronauts available to command future flights. Senior astronauts assumed, without ever articulating it, that they would never serve in a subordinate role to any astronaut selected after them; that having received a command seat, they would never be demoted; and that members of the Original Seven would always lead any crew to which they were assigned. To command each mission, Slayton assigned a remaining member of the Original Seven (if available), or a trusted member of the 1962 selection. Some missions might even be so demanding as to require two or more experienced men; other seats, though, were best filled by promising rookies who might command future missions once they had acquired sufficient experience in space. In this way, pairs of astronauts were joined in tentative crews: teams that served first as backups to prime crews before flying their own missions.

Slayton expended little effort to match crews for personal compatibility; he assumed that such driven men would "get along no matter how they were matched up" and only occasionally matched friends on missions to make "life easier" or add "some fun."[82] "I think there's some idea with the public that NASA must run us all through computers according to our personalities so that we all come out matching," noted astronaut Ed Gibson in a 1973 interview. "That's not true. . . . On later missions, an intense work schedule kept the men from socializing...or fighting."[83] Slayton expected astronauts to sort out their differences and work toward their shared goal. "[If] Deke said that he was going to put you in there," one junior astronaut later noted, "you'd damn well better get along or your ass was out of there!"[84]

As the training required demanded that a prime crew and backup crews be assembled well in advance of each flight, crew rosters became a critical measure of an astronaut's future in the organization. Backup duty was a common source of frustration among crews: astronauts assigned to it trained for the mission as if they were going to fly it themselves, waiting for the calamity that might enable them to replace the members of the prime crew. For junior astronauts in particular, though, assignment to a backup crew early in flight schedule of a new vehicle virtually guaranteed a plum prime crew assignment three missions later, when the backup crew rotated into the prime crew position. By contrast, an assignment to backup the final missions of a project ensured that the crew would not only miss an opportunity to fly the vehicle for which they trained, but would be too consumed with backup duties to prepare for flights on the next new vehicle. (Occasionally, these backup positions were filled by newly returned, "dead-end" crews expecting no further flying opportunities in a particular program.)

Ostensibly, a variety of factors, including academic training, management experience, peer ratings, and performance determined who earned the right to fly and when.[85] In practice, though, Slayton's first impressions swamped any formal decision-making tools; astronauts were often unable to decipher Slayton's reasoning and, in any event, saw little value in challenging his assessments. In late 1964, Slayton mysteriously asked the astronauts of the 1963 group to evaluate each other's flight-worthiness, a management technique Gilruth had earlier employed with the Original Seven.[86] Slayton later admitted that the peer ratings were Gilruth's idea and that Slayton hadn't "paid a lot of attention to the results," as he had already picked the flight rotation the previous year. Brutally honest about their colleagues, the astronauts, in their peer ratings, virtually duplicated Slayton's rotation, yet the rookies remained convinced that an arbitrary formula known only to Slayton determined their future.[87]

Slayton's attitude toward academic credentialing proved the greatest source of consternation to some of the better-educated junior astronauts. While men might study spaceflight in lecture halls and classrooms, Slayton intimated, such study did not make them fit for flight. Like Robert Gilruth, Slayton had received a BS in aeronautical engineering from the University of Minnesota, but Slayton dismissed higher education as an affectation unrelated to one's success in space navigation. "People have the mistaken idea that your education or professional training is what you call on during a Gemini or Apollo mission," Slayton later wrote. Rather, Slayton concluded, formal academic training was of little use in flying spacecraft, a conclusion he quickly reached despite having no space piloting experience himself. Individuals who attempted to leverage their academic qualifications to obtain a crew assignments were rebuffed; Slayton was unimpressed by university credentials and even denigrated Aldrin's ScD dissertation on orbital rendezvous, for example, describing it erroneously as a PhD and noting that he declined to assign Aldrin to an early orbital rendezvous flight despite his extraordinary familiarity with the subject.[88] Slayton's dismissal of book learning, though, was inconsistent; at the same time he denigrated higher education, Slayton ruthlessly promoted formal test pilot training as essential for advancement in the Astronaut Corps. It is unclear, though, if Slayton valued test pilots for what they knew or for the character traits he thought test pilots inherently possessed: courage and reliability far in excess of that possessed even by other pilots.

Exactly which junior astronauts met Slayton's standards was apparent from the assignment logs: test pilots—especially those personally known to senior astronauts—most often received the best opportunities. Those viewed as especially competent were fast-tracked for piloting duties and early command, while others languished, denied flight opportunities or finding themselves in crew positions with little growth potential.[89] Upon the arrival of the 1963 selections, for example, Slayton selected the test pilots for Gemini flights and training as Apollo CMP, among the demanding and critical positions in the upcoming missions and a sure path to promotion.[90] Nontest pilots, especially those with scientific training, were received warily and could do little to redeem themselves.[91] Several members of the selection—Aldrin, Cunningham, Chaffee, Bill Anders—were assigned to be Lunar Module Pilots (LMP), a less demanding crew position in which the chance to walk on the Moon balanced the diminished chances of later promotion.[92]

The rankings struck many junior astronauts as inherently unfair. To those, like Cunningham, on the back end of Slayton's list, his seemingly magical guidelines were "astropolitics." Rather than rewarding talent, the "rules" were a kind of cronyism that equated tenure in NASA, popularity with superiors, military pedigree, and flying experience with astronautical skill. To Cunningham, whose status as an operational fighter pilot with an advanced degree doomed him to low-status crew positions, Slayton's hierarchy produced the seemingly perverse result that some astronauts might visit the Moon twice while others never got near it. Indeed, because their experience was so valuable, astronauts who had already flown in the subordinate role of CMP would likely be invited to command an Apollo lunar flight, where the experience would come in handy. Thus, securing an early prime crew lunar mission seat actually increased one's chances of securing another, even with competition from large numbers of able rookies.[93]

The worst element of astropolitics, though, Cunningham suggested, was that personal relationships seemed to matter more than competence; junior astronauts who knew veterans through military work or past missions could find themselves on future flights next to their mentors.[94] Several such crews—Frank Borman and Jim Lovell, Tom Stafford and Gene Cernan—often returned to space intact, limiting opportunities for newcomers.[95] Stafford reputedly justified Cernan's frequent pairing with him by noting that Cernan "does exactly what I tell him."[96] Junior astronauts without friends, though, faced additional challenges. The humorless and often irascible Borman, for example, nearly became the first in his class to fly until Gemini 3 commander Grissom booted him from the flight, apparently due to personality clashes.[97]

Seasoned commanders accumulated political "capital" within NASA that they could use to claim future missions for themselves or their protégés.[98] When Stafford turned down command of Apollo 17, his lobbying helped to ensure that Cernan would be offered the position.[99] In cultivating their protégés, astronauts acted as teachers and, in some cases, parents, chastising their charges like children. Stafford reputedly lobbied Slayton for Cernan to receive his Apollo 17 command, making him the only LMP to receive such a promotion. To his chagrin, though, Stafford's efforts were almost undermined by Cernan's crashing of a training helicopter weeks before the flight. "You dumb fucking shit, what the hell did you do?" Cernan recounted Slayton chastising him on the telephone after the accident. "I'm out here trying to make sure you command Seventeen, and you just may have fucked it all up."[100] More often than not, the veterans had gravitated to certain rookies out of respect for their abilities, but to outsiders, the affection they showed toward their protégés was unseemly.

Socially skilled and extremely competitive, less well-connected astronauts lobbied, pressured, and manipulated each other for the privilege of flying, often dismissing or denigrating colleagues to improve their own standing. "If you can't say anything good about someone," Cunningham declared, "don't hesitate."[101] Astronauts labored under constant scrutiny, hoping to avoid goofs that would undermine Slayton's fragile confidence in them. Minor mistakes, a hint of disobedience, or a bout of space sickness on a previous mission could ground a promising astronaut permanently.[102] Insubordination by commanders was imparted to their crews, and could keep them from future flights.[103] Junior astronauts' technical assignments or scientific interests might undermine confidence in them as pilots or tie them to future projects at the expense of immediate flying opportunities. Even seeming promotions—to command a backup crew or head the office of a follow-on program—could be detrimental to one's flying career if that mission was the last in a series or the office without prestige. As a result of his performance on Apollo 8, LMP Anders was promoted to CMP, an honor that made future command a possibility but robbed him of a chance to walk on the Moon in the immediate future.[104]

Trading opinions on how each of them might earn a prime crew slot, Cunningham and his colleagues debated and gossiped, but to no avail: Cunningham's "useless intelligence network" of bottom-tier astronauts could not plumb the depths of Slayton's reasoning or change his mind.[105] While some of the astronauts seemed to have arrived at NASA already so respected that they could do little to undermine the momentum that catapulted them to early leadership positions, others seemed almost doomed to disappear no matter what they said or did. Cunningham's classmate Aldrin, who combined scientific preoccupations with a "competitive" and "direct" personality, found himself unprepared for the subtle office dynamics of NASA. Aldrin talked incessantly about orbital mechanics, insinuated himself into mission planning uninvited, and generally annoyed people. "I just wasn't an organization man," Aldrin later recounted; the subtle self-promotion of the astronaut corps had to be backed by the proper piloting pedigree and the kind of gentle politicking of which Aldrin was incapable. "I was really an odd man out" Aldrin recalled, "…they didn't really want somebody who hadn't been in the test pilot business."[106] Shunted at first to a dead-end backup slot, Aldrin eventually flew on the last Gemini flight, Gemini XII, where his unique skill at orbital mechanics proved critical when the craft's radar failed.

Other rookies, though, hoped that by quietly pursuing their work and making no waves, they would impress veterans without resorting to the kind of vigorous attention-getting that others attempted. Virtually all of the LMPs—Anders, Bean, Chaffee, Cunningham, and Eisele—applied this strategy with depressing results; instead of being impressed, Slayton simply ignored the men, assuming they were weaker than their peers and assigning certain of them to early missions unlikely to tax their abilities.[107] "…I can't say that I was a good contributor," Bean recalled in 1998. "I wanted to be every day, but I never figured it out. …[I]nteracting with other astronauts, I wasn't very good at that."[108] "…I would go present this conclusion to Al Shepard or Deke Slayton, they weren't interested and they thought I was a little nutty." "I can remember at those times how we used to have so many conversations about, 'If I were Deke, I'd do this,' or, 'If I were head of NASA, I'd do this,' " Bean recalled. "What the hell did we know? Nothing. We were barely able to fly a spaceship. And not only that, no one was asking us… ."[109]

First Flight

Sharing hotel rooms on the road, Cunningham and his friends "bitched about crew selections" and the "nebulous higher authority" holding them back—until they received a crew assignment.[110] The best morale- and career-boost for an astronaut was a flight assignment followed by successful performance in space; Aldrin's work on Gemini XII earned him a spot on Apollo 11, while Bean received a later Apollo command after serving as LMP on Apollo 12.[111] "Once a man was assigned to a flight," Ruff and Korchin noted with some degree of understatement, "further delays had little impact" on morale.[112]

For most junior astronauts, assignment to a backup crew was the first step in their preparation for actual flight, and a chance to replace some of their more tedious engineering duties with flying work.[113] "You see the conveyor belt moving," Bean recalls of his assignment to backup Apollo 9 and then fly in Apollo 12. "[Y]our friends are flying every two months, so you see these things happening and you know that you just have to stay healthy and concentrate, and then you'll get your chance."[114] From the moment of assignment, the astronaut's world narrowed considerably, training intensified, and concerns shifted toward anything that might jeopardize the flight. Missions were highly choreographed, and to prepare for them, astronauts assigned to crews would spend years training with the relevant technology and practicing necessary procedures, often using sophisticated vehicle simulators. (Indeed, simulator work eventually came to occupy half the astronauts' time in training.[115]) To observers, NASA's public image during the 1960s was one of technical competence bordering on infallibility, but what characterized astronauts' preparation was a carefully calibrated expectation of failure. Astronauts trained for success, but also for innumerable problems, rehearsing corrective maneuvers, planning for catastrophe, and anticipating trouble. Bean, assigned to backup John Young's Gemini X crew, was relieved, ultimately, that he did not have to fly the mission himself: as an astronaut, Bean feared, he wasn't very good: "we wouldn't have had the flexibility and the understanding" to handle in-flight mishaps properly.[116]

Astronauts participated in every facet of each other's flight, as backup or support crew members, by supervising hardware design, or by contributing in any number of informal ways. Indeed, through the flight, other astronauts served as the flight crew's only link to the outside world, waking them, preparing them, communicating with them. From the Project Mercury onward, a fellow astronaut typically joined the engineers of the "closeout crew" in seating the astronauts and readying them for launch.[117] Perpetuating the idea that no one could understand the astronauts' job better than the astronauts themselves, astronauts wrote their own flight checklists and occupied ground control stations as Capsule Communicators, insuring that messages to astronauts would be communicated only by fellow astronauts, a practice borrowed from the X-15 program.[118] And, like a doctor pronouncing death, only an astronaut could announce when a launch vehicle had properly lifted off, by observing it on a TV monitor.[119]

With luck, a backup assignment ended with a seat on a prime crew. Reflecting on his preparation for his first spaceflight, the 1966 Gemini X mission, Michael Collins could not say how long he had trained, estimating that he had devoted his entire three-year career at NASA to the task:

> For six months I had been training specifically for Gemini 10, with its peculiar EVA and rendezvous problems, but this period had of course been preceded by six months of Gemini basic during my tenure as backup for Gemini 7. ...But how about all the Basic Grubby Training, with its emphasis on orbital mechanics, or the jungle training, or centrifuge, or zero-G airplane, or pressure suit work? Or what I learned test flying at Edwards, or ejecting from a smoke-filled cockpit years before? Or the math I learned in school, the foundation upon which orbital mechanics was built? I just don't know how long it takes, except that it has taken me thirty-five years to reach crew quarters... ."[120]

Collins had trained as intensely for his second flight, Apollo 11, but, even so, felt he needed more. Apollo's training plan called for between 200 and 250 hours of simulator time in addition to other preparations; Collins thought 400 hours should be the minimum—twice the length of the actual flight. "I tried to learn all the systems; I tried to learn all the procedures for burns, all the rendezvous procedures, and the navigation," Collins recounted upon his return, "but I will be the first to admit I was far from being an expert in any one of these fields."[121] Yet, Collins's familiarity with Apollo equipment, like his spacesuit, was extensive; Collins recounted in his mission debriefing that, "[a]lmost every day for 3 months before flight, we were in that suit sometime during the day."[122]

From the moment space-bound astronauts woke up, theirs was a world of almost infinite schedules and procedures; even the breakfast they ate before flight had been specially concocted by NASA nutritionists to ensure that they would not defecate in their suits, the most intractable excretory problem of human spaceflight, and one that has yet to be conquered.[123] With backup crews always at the ready, replacement due to illness or injury was a constant threat right through launch, but once an astronaut's launch day arrived, his thoughts could be remarkably mundane. For Collins, it was the fact that he had to get up much too early in the morning for his liking: the motion of the Earth and the Moon required launch at specific times, and for Apollo 11, the launch "window" opened at 8:32 am, requiring Slayton to wake the crew "shortly after four o'clock."[124]

Walking along the tower gantry to their vehicle, Ed Gibson recalled of his 1973 launch, he could hear the "creaking and groaning" and the rocket's metal skin "shrinking" from the cold of the propellants inside of it. Inside the spacecraft, lights illuminated the machine which, after months of preparation, had "started to come alive." "And then. . .all of a sudden the bottom floor of the building explodes." The feeling of the rocket's ignition was one of "turbulence" rather than motion, with violent oscillations like "being a fly glued to a paint shaker." Jolts continued, as the rocket's first stage expired and its second stage lit, accelerating the astronauts and pressing them into their seats with four times their own weight. Once in orbit, those motors quit as well, leaving the astronauts weightless, and their space vehicle filled with debris that on Earth had been safely tucked in corners (until the cabin ventilation systems filtered them out).[125] Out of the windows, familiar continents would begin to roll by.

Nineteenth-century physicians, Wolfgang Schivelbusch wrote in *The Railway Journey*, feared that locomotive transportation would subject riders to bone-shattering shocks and organ displacements.[126] Aviation physicians of the 1950s had similar concerns about spaceflight, wondering if the human body would still be able to pump blood and digest food amid the weightless conditions of orbit. "Human factors" research was a principal goal of Project Mercury, and mission design decisions weighed heavily on expectations of crew performance under stressful conditions. The final configuration of Apollo, for example, was influenced by fears that disoriented astronauts, landing a vehicle on the Moon while lying on their backs looking upward (less-than-ideal even for terrestrial aviators) might lead to a lunar crash if attempted in space.[127] While early spaceflight bore out none of the physiologists' worst fears, the experience was profoundly uncomfortable, a fact astronauts tended to conceal so as to prevent any suggestion of physical weaknesses that might jeopardize future flight assignments. In orbit, the crew of Gemini X struggled with a host of medical issues that they had neglected to report to ground personnel: Collins with claustrophobia and the bends, and both he and Young with eye irritation, prompting Young to fear (only partly in jest) that NASA managers might call him a "sissy" if he reported "crying all night." [128] "Discuss it or ignore it?" Collins further debated about his knee pain:

> I have a vivid picture of the avalanche of medical conferences one quick complaint will produce. It will cause everything short of a house call. At a minimum, it will result in a stream of excited radio transmissions lasting half the night. This I don't need.[129]

Frank Borman, commanding the first circumnavigation of the Moon in 1968's Apollo 8 mission, was incapacitated for several hours early in the mission from fever, vomiting, and diarrhea, ailments he avoided disclosing to ground physicians, who, when they found out, debated aborting the mission, fearing a crew-wide viral infection.[130]

For even the most healthy spacefarers, blood pooled in places where gravity would normally keep it from collecting (like the head), and vestibular effects eliminated all sense of orientation.[131] At first, these effects were more uncomfortable than debilitating. Later astronauts contended with larger, multicrewed craft that posed new issues of health and comfort. To combat muscle atrophy on long flights, astronauts took to exercise, with minimal effect.[132] The most significant health issue was space adaptation syndrome (or "space sickness"), an inner ear disorder that produced, in a certain percentage of people, stomach awareness, nausea, and often vomiting for the first 24 hours of flight. During Mercury and Gemini, though, flights were so short and vehicles so small that astronauts could seldom move enough to feel any disorientation. During Apollo 8, space sickness became a common and potentially debilitating ailment, even for experienced astronaut Borman, the mission's commander.[133] In the larger cabins of later vehicles like Skylab and the space shuttle, astronauts struggled even more with weightlessness, tumbling, flailing, and often attempting to swim, unsuccessfully, through the air, a medium lacking the density for such propulsion.[134] That mission control would communicate constantly with crew members, soothing their spirits and providing helpful information was another popular misconception; while later Apollo crews enjoyed near 24-hour communications, Apollo 7's crew was in range of ground stations only 5 percent of the time.[135]

Even absent the nausea and vomiting of space sickness, eating, excreting, sleeping, stretching out and maintaining one's body temperature in the cramped, weightless confines of a space capsule proved intermittently problematic. Claustrophobic early space vehicles offered unpalatable food and reeked of urine and body odor.[136] (Indeed, Gemini vehicles intended for two-week-long operation offered even less room per person than the closet-like Mercury spacecraft.) Mercury's short flights had featured pureed meals crammed into toothpaste tubes. Bite-sized Project Gemini foodstuffs were engineered so they would not crumble or melt, and then covered in wax to preserve them; they were often unpalatable, especially after several days.[137] Other nourishment was provided through a variety of irradiated, canned, or dehydrated powders mixed with water supplied by the spacecraft's fuel cells. In the weightless environment, gas bubbles dissolved in liquids did not vent to the surface, leading to painful flatulence.[138]

While food improved with project Apollo, hygiene did not.[139] Waste management posed two critical issues: the first was how to manage excreted material without violating suit integrity; the second was the danger of noxious debris floating throughout the vehicle in the weightlessness of orbit. Project Mercury engineers resolved these problems first by ignoring them, and then by making only limited provisions for urine collection and none for solid waste. Astronauts beginning with Grissom wore an external catheter, a modified condom-and-tube apparatus draining into a bag strapped to their leg; astronauts were also started on a low-residue diet prior to lift-off.[140] Waste collection was more than an issue of modesty; on Cooper's day-long 1963 Mercury flight, the urine collection apparatus failed, leaking fluid and damaging the reentry guidance electronics.[141] For Mercury's short flights, lack of a formal waste management system was a minimal inconvenience, but engineers could not ignore it for the week-long voyages in more complex vehicles sure to follow. Apollo guidance engineers eventually sealed the spacecrafts' electronics to keep astronauts from inadvertently urinating on them, and biomedical experts reexamined the solid waste question.[142] One solution was pharmacological; astronauts flying the two-man Project Gemini flights of 1965 and 1966 were prescribed Pfizer's antidiarrheal LOMOTIL® (diphenoxylate and atropine) to inhibit defecation.[143] On these and later flights, drugs were supplemented with "messy" antiseptic fecal collection bags and wipes that required the astronaut to devote nearly an hour to each bowel movement.[144] Moonwalking astronauts eventually relied upon a diaper system to ensure "comfort" during their multihour sojourns on the lunar surface. Sexual health was a nonconcern; even on his record-breaking 84-day Earth orbital mission, Ed Gibson reported neglecting that aspect of his physiology completely.[145]

Though well-prepared, each astronaut discovered for himself a manifold of details while in space, learning where to wedge one's head to emulate the feel of a pillow in a weightless cabin, or how to hold one's hands while sleeping so as not to accidentally trigger retrofire.[146] In "free fall" during their entire voyage, most sleeping astronauts were consumed by the terrifying sensation of jumping off a diving board and never reaching the water.[147] In space, the sun rose and set once each every 90 minutes, but astronauts kept Houston time and tried to maintain a normal schedule, made more difficult by the fact that missions corresponded to eastern time (Cape Kennedy) instead of central time (Houston).[148] NASA's Space Task Group had settled upon a three-person Apollo crew in the expectation that they would sleep in shifts, but this eventually proved impossible, and the first Apollo crew got little sleep, annoying each other and leading one crew member to fall asleep while on watch.[149]

Piloting such vehicles was often challenging, and became more so, as the space program progressed. The earliest spacecraft had flown automatically, but the influence of NASA managers from the flight test world, David Mindell writes, prevented the advocates of automated systems from writing humans completely "out of the loop."[150] Astronauts, having accepted that they could not fly their rockets "off the pad," accepted their new role as systems managers who would fly their craft aided by automated devices of various kinds, represented by 100+ switches on Mercury and Gemini vehicles, and over 400 in the Apollo Command Module.[151] Rather than freeing astronauts from responsibilities, new computer technologies and cockpit instrumentation in Gemini and Apollo spacecraft increased the workload of astronauts, who needed to become proficient systems engineers if they were to remain in control of their craft.[152] For designers, this meant adding human controls to automated missile guidance systems that did not previously require them; for astronauts, learning to fly airplanes in an environment without air.

In space, astronauts often found their terrestrial piloting experience less than useful. Orbital mechanics, the dragless vacuum of space, and the zero-g environment of orbit confounded the pilots' intuition about the effects of force and movement, leaving one astronaut confused about a "mysterious force" frustrating his attempts to control his body and manipulate objects.[153] (In fact, the mysterious force was merely Isaac Newton's Third Law of Motion, pushing him away from every object he touched.) In orbit, even the most conventional piloting wisdom—speeding to catch a distant object—didn't work. Accelerating in orbit, Jim McDivitt found on Gemini 4, only raised his orbital altitude, pulling him further away. Rather, to speed up, one had to slow down, lowering the orbital altitude and thus increasing the speed.[154] Mike Collins and his air force test pilots at Edwards had calculated the physics in preparation for future adventures, but once in space, astronauts, limited by their terrestrial instincts, encountered difficulties putting orbital mechanics into practice.[155]

Mercury spacecraft offered limited control opportunities; small and designed to handle like a fighter jet, the Gemini capsules were the astronauts' favorite.[156] Nudging hand controllers, astronauts could maneuver their tiny vessels around each other in orbit, aided by a digital electronic computer but not beholden to it. Apollo lacked Gemini's nimble handling but possessed more robust computing power that required more intensive training.[157] Open cockpit barnstorming this was not: rarely in Apollo flights did astronauts assume the raw piloting responsibilities the public imagined of them. Recounts David Mindell in *Digital Apollo*:

> For the astronauts, an Apollo launch was a ride atop a fiery automaton, as they watched dials and indicators, poised for an abort while the rocket executed its sequence of staging, steering, and burns. They spent their trip to the moon largely doing systems monitoring, maintenance, and housekeeping, complemented by star sightings to back up navigation from earth. Major rocket burns were calculated in advance by computer and the ground controllers, directed and controlled by servos. Only when approaching the lunar surface would

> pilots do what they did best: fly and land a delicate, powerful craft. Yet even there, only the last minute or two of the ten-minute descent would be under manual control. Here "manual" meant jogging a stick that would provide new setpoints to computer-controlled feedback loops, either for attitude holds or descent rates.[158]

To American industrial workers, the arrival of mass production machinery on the factory floor served as a new, troubling form of labor discipline, but early astronauts welcomed the arrival of their complex new "spacecraft," confident that they could operate these vehicles better than anyone else on the planet. Like the highly skilled metal workers who struggled to retain control of their machines despite the arrival of computer-controlled machine tools, astronauts recognized that success in the space program required an almost machine-like precision in the operation of unforgiving high-performance vehicles.[159] Rather than spurn these difficult craft for less demanding automated ones, astronauts welcomed them, in the hope that such craft were less apt to fail, would preserve their role in the program, and would discourage the recruitment of astronauts less skilled than them.

Despite the risks of their environment, though, the astronauts' work could be remarkably routine. Junior crew members took their cues from their commander, while the commander followed the orders of the flight director, Chris Kraft. Too busy to sightsee, astronauts, at least during the early portions of their flights, concentrated on the vehicles' complex systems. Apollo 8 LMP Bill Anders recalled that Commander Frank Borman warned him not to look out the window; "I peeked a couple of times, but he never caught me."[160] When in sight of ground stations (and not sleeping), later Apollo astronauts were in constant communication with colleagues on the ground, receiving mission updates and diagnosing technical problems. Experience on a previous mission, Collins noted, though, had helped him prepare for the unusual sensations and discomforts of his Apollo flight. "I had been up there in zero g before and I wasn't spending all my time pondering the wonder of it all."[161]

Complex flight plans and elaborate checklists—what Collins called Apollo's "fourth crewmember"[162]—prescribed crew activities for nearly every second of the day and discouraged sightseeing. Aware of the complexity of the space environment, astronauts brought the technology of the checklist from the flight test world and made it central to NASA's human spaceflight efforts. Creating, revising, and supervising their production, astronauts relied upon these printed lists and loose-leaf notebooks—some over 100 pages long—to reduce the burden on their memories, increase the speed of operations, and prevent life-threatening mistakes in cockpits littered with hundred of switches, knobs, and dials. As the astronauts' piloting skills became less crucial, their skill and familiarity with checklists became part of their professional toolkit, providing them with a new kind of work product that they, as the most expert operators of the vehicles they flew, were best able to prepare.[163]

Astronauts' communications with the Mission Control Center about these documents alternated between highly technical discussions of vehicle systems and flight activities and the kind of banter that could have been found on any office telephone system in America:

07 13 26 46	CMP	Houston, Apollo 11. Over.
07 13 26 55	CC	Apollo 11, Houston. Go ahead.
07 13 27 02	CMP	Roger, Houston. For RETRO, I have the anticipated location of all the entry stowage, and I suggest you pull out the entry checklist, and we'll go through those maps in the front of it.
07 13 27 19	CC	Apollo 11, Houston. Could you stand by just a few minutes? Charlie and Flight are out getting a weather briefing. They'll be back shortly.
07 13 27 33	CMP	Is this Ken?
07 13 27 35	CC	Say again?
07 13 27 40	CMP	Is this Owen?
07 13 27 42	CC	No, this is Chuck Lewis. Charlie Duke is out with Flight getting a weather briefing right now.
07 13 27 49	CMP	Okay. They're out drinking coffee. I know.
07 13 27 52	CC	(Laughter) They'll be back momentarily.[164]

Often, ground engineers and astronauts attempted to resolve mechanical problems, working as an engineering team, teleconferencing over tens of thousands of miles:

06 05 03 20	CC	Roger. If Neil has a free minute, we've got a question or two regarding the CO2 partial pressure and water in the suit loop discrepancies noted yesterday. Over.
06 05 03 33	CDR	Go ahead.
06 05 03 36	CC	Roger, 11. Was water noted in both suits or only in yours, Neil?
06 05 03 44	CDR	I think only in my suit.

* * *

06 05 05 44	LMP	I should mention, Bruce, that when I went to water secondary—water separator to secondary there, I didn't notice any change. But about after 15 minutes or 20 minutes, the water stopped coming out. So maybe that was just water that was already in the loop that was still blowing out, but the secondary water separator was operating properly.
06 05 06 11	CC	Roger. Did you make any changes in the suit loop configuration after you went from the egress mode to the cabin mode after insertion; that is, in particular, they're interested in knowing if you recall changing the diverter valve position to EGRESS at any time while you were on the secondary canister? Over.

06 05 06 36	CMP	No. I don't believe we did that at all, Bruce.
06 05 06 40	CC	Okay, 11. Thank you. That sums up our questions for now, and we'll crank these back into the engineering pipeline and see what we can come up with.[165]

Rarely did serious problems affect the mission, but innumerable minor systems failures—stuck hatches, faulty cooling systems, inadequate hygiene equipment—were constant and often made the astronauts' stay in space uncomfortable. Frequently, these issues required "work-arounds"—alterations to standard procedures—and manual manipulation of hardware, validating NASA's decision to put such well-trained personnel aboard its vehicles. Encountering faulty instrumentation minutes into the Gemini 3 flight, John Young coolly diagnosed the problem and switched to a backup electrical system;[166] unable to shut the hatch after Ed White reentered their Gemini IV spacecraft, Jim McDivitt finessed an errant cog with his gloved hand.[167] Seeking to ensure that the public remain convinced that its spaceflight ambitions were feasible, though, NASA seldom publicized these small problems. Only infrequently did a problem force an astronaut to manually pilot his spacecraft or make a critical, life-threatening decision, but it was the ever-present possibility that such interventions would be necessary that enabled the astronauts to justify their presence in space.

The longer a spacecraft remained in orbit the more of its hardware tended to break down, but during Project Gemini's long duration flights (which put crews in space for up to 14 days), boredom proved as potent an annoyance as mechanical failure. On later Apollo flights, overpacked schedules in orbit left astronauts scrambling for free time to rest, look out of the window, and float in the cabin, but early flights posed the opposite problem: intended merely to test vital engineering systems, missions lacked full work schedules or diversionary activities.[168] NASA, initially, engaged in little planning for astronaut recreation, leading to a profusion of stunts and inside jokes between crewmembers and ground controllers, many involving props—a corned beef sandwich, jingle bells, a harmonica, an array of signs—smuggled aboard spacecraft. Without such distractions, long test flights could be monotonous, especially when, in an effort to conserve systems and vital thruster propellant, astronauts refrained from maneuvering their spacecraft and allowed them to slowly tumble in orbit.[169] Without the amusement of piloting, space travel effectively lost its glamour. "I call it eight days in a garbage can," Pete Conrad recalled. "The fact is, you can't do anything. You can't go anywhere. You can't move and have no great desire to sleep because you're not doing anything to make you tired. You don't have anything to read and there isn't any music. ...I spent half my life opening stuff up and re-wrapping it...so we didn't have garbage all over the cockpit. If it wasn't for that I would have probably shot myself... ."[170]

For most astronauts, any relief over such a mission's conclusion was tempered by the fact that a spacecraft's return to Earth was only slightly less dangerous than its launch. In Earth orbital flights, a critical firing of the spacecraft's rocket motors against the vehicle's direction of movement slowed the spacecraft enough for its orbit to decay, causing the capsule to descend into the atmosphere at hypersonic speed. At too shallow an angle, the capsule would skip off the atmosphere and back into orbit; too much, and it would plunge so quickly that friction from the air would vaporize the craft. Returning from the Moon, astronauts faced a plunge toward the Earth for which they had only one chance and little room for error. Before flight, psychiatrists noted, astronauts performed slightly above baseline on various tests of their mental faculties; after flight, astronauts generally demonstrated heightened awareness and mental functioning.[171] Physically, though, they were often weak and unstable, unused to balancing their weight, and as one later astronaut recalled, unable to walk through doorways without bumping into them.[172]

Despite the opportunities for contemplation seemingly afforded by their voyages, the earliest space vehicles were the sites of few tantrums, epiphanies, or philosophical debates. Surrounded by dangerous technologies, most astronauts were subdued and business-like in flight. Despite the frequent discomforts and occasional surprises of orbital flight, few of the men admitted to being paralyzed with fear while hurtling through space; all, George Ruff and Sheldon Korchin concluded, exhibited a "conscious control of emotion arising from external danger." Falling back on "past experience in the mastery of stress," their "training and technical readiness," their "ego strength" and "self-esteem," astronauts endured life-threatening crises with a minimum of emotion, concealing any hint of fear or physical discomfort.[173]

If Tom Wolfe later articulated a culture of pathological risk-taking among the earliest astronauts, Norman Mailer, years earlier, described certain Apollo astronauts as interesting largely for their near total lack of visible affect. A few of America's space heroes, Mailer writes, were sphinx-like and impenetrable even to their own families, rarely excited, seldom given to self-expression, and at times, distant, even when off the clock.[174] Buzz Aldrin, LMP on Apollo 11, admitted rarely ever engaging in "free exchanges of sentiment" with Commander Neil Armstrong even during months of pre-flight training, much of which had been spent standing side-by-side in a compartment the size of a small bathroom.[175] Though frequently likening themselves to soldiers, the astronauts, NASA psychiatrists noted, did not form the kind of intense personal bonds that characterize "infantry squads," building a sense of community instead upon "common professional values" centering on "respect for technical competence." Assessing the astronauts in 1964, Korchin and Ruff noted an absence of "deep attachments" among a group that seemed to rely upon each other for their very lives. "It is faith in the expertise of the man, rather than dependence on the man himself, that allows them to accept interdependence without suspicion."[176]

Nor were NASA or the early astronauts inclined to complicate the space program with philosophical debate, artistic reflection, or insincere religiosity.[177] Indeed, even brief efforts by certain astronauts to incorporate religious practice into missions created controversy for an agency aware that Congressional oversight committees were reluctant to spend public funds on interplanetary missionary work. When invited to engage in spiritual or philosophical discussions, musings on the meaning of space travel eluded the astronauts. Gene Cernan, Commander of Apollo 17, recalled his feelings of inadequacy at being unable to "give people the answer they want" when asked about walking on the Moon.[178] John Young, the puckish astronaut who smuggled a corned beef sandwich aboard Gemini 3 and later complained about flatulence on the lunar surface was, remarked Gemini X copilot Collins, the most "uncommunicative" astronaut with whom he had worked. "I don't have any idea what flying in space has meant, or will mean to him," noted Collins.[179]

Even more respected by NASA than machine-like reliability was imperturbability in the face of mechanical failure. Such resourcefulness validated human piloting and advertised the unique skills present with the burgeoning Astronaut Corps. Indeed, writing in 1964 on the results of the Mercury flights, Ruff and Korchin noted that for one astronaut they interviewed, the "failure of automatic instruments intensified the feelings of personal contribution and success in the mission."[180] During an aborted launch in 1966, Schirra refused to allow himself to be rattled when his Titan II launch vehicle failed to budge; instead of pulling the handle that would eject him and pilot Stafford from their Gemini capsule, he stayed put, saving the spacecraft and the mission, which launched successfully three days later.[181] (In recognition of his bravery, Schirra eventually received a medal, literally, for doing nothing.) In 1966, Armstrong recovered from a potentially lethal in-flight mishap on the Gemini VIII flight, summoning up enough strength and coordination to end an uncontrollable spin threatening to leave him and his crewmate, Scott, unconscious.[182] And Cernan recalled finding himself, during the launch of Apollo 17 in 1972, almost wishing that the automatic guidance system would fail, giving him the opportunity to gimbal the Saturn V launch vehicle's engines manually. "I almost dared her to quit on me," Cernan later recounted.[183]

Ballad of the Bad Astronaut

In *Space and the American Imagination*, Howard McCurdy describes how images of the imperturbable, reliable astronaut helped to establish the "aura of competence" surrounding spaceflight during the Apollo era. By appearing to be an organization that could get things done, NASA increased popular confidence in both the space program and the government as a whole, a powerful argument for the space program's continuation through the mid-1960s, despite its high cost.[184] Astronauts figured prominently in these calculations; selfless and skilled as aviators, they personified the competence of the space program and so thoroughly represented American values—courage, service, faithfulness—that any failure on their part would have reflected poorly on the nation that produced them.[185] NASA's astronauts, though, mostly wished only to succeed: to control their fear, avoid errors, and return themselves and their craft home in reasonable condition.

Despite official statements to the contrary, not every astronaut performed equally in the cockpit, and even the best made mistakes. NASA, in particular, seemed to turn a blind eye when respected pilots wrecked training aircraft, a fairly routine occurrence. Having received relatively little training in NASA-supplied helicopters, the astronauts nonetheless flew them with the false confidence of veteran pilots. Astronaut Joe Engle ran short of fuel before landing; he had brought the helicopter down and survived uninjured, but destroyed the aircraft. His crash-landing—using a technique for which he had never trained—had been so skillful that no one criticized him.[186] Neil Armstrong also seemed able to survive a catastrophic mishap with his reputation actually burnished. In May 1968, Armstrong was forced to eject from a jet-powered lunar landing simulator less than a second before it hit ground near Ellington Field. The jet-powered Lunar Landing Test Vehicle relied upon rocket thrusters for steering, and Armstrong had unwittingly run the tanks dry. NASA fudged the press release to downplay the near-catastrophe, accident investigators blamed leaky thrusters, and engineers blamed high winds for leading Armstrong to exhaust his fuel sooner than usual. To the surprise of his colleagues, though, Armstrong was back at his desk, quietly working, an hour after ejecting, a lack of affect that Alan Bean regarded as odd, even for an astronaut.[187]

Confronted by personal failure, the best astronauts stewed and obsessed, but always kept working. After using too much fuel during a docking maneuver on 1966's Gemini X flight, John Young sulked in the cockpit for the rest of the day as he set about his work.[188] Accidentally erasing navigation data in Apollo 8's computer, Jim Lovell and the rest of the crew wasted no time arguing, setting to work attempting to reorient the spacecraft.[189] Buying in to their own mythology, astronauts often consoled themselves with the idea that mistakes were the fault of others: an astronaut who made a simple error might blame poorly designed hardware. Occasionally, divergent ideas about technology between the people who had created them—scientists vs. engineers, astronauts vs. ground personnel—could produce problems in flight. A series of computer overloads in the Apollo 11 Lunar Module, astronaut Buzz Aldrin later recounted, owed to a failure of engineers to appreciate the demands of actual flight, leading to them approving procedures that led Aldrin to overtax the computer's limited computational capacity.[190]

Walter Cunningham expected that a record of in-flight errors and bizarre crashes would establish one as a poor choice for a prime crew seat, but this did not appear to be the case. Rather, the Astronaut Office seemed to expect that its aggressive pilots would occasionally break their aircraft, recognizing in-flight mishaps of all kinds as the inevitable for aggressive aviators. In the process, though, NASA turned a blind eye to flying habits and behaviors that were needlessly reckless. Gus Grissom lost his spacecraft shortly after splashdown, yet he retained the confidence of his superiors. Sitting in his capsule off the coast of Florida and awaiting recovery by a helicopter, Grissom was startled by the premature detonation of explosive bolts sealing his escape hatch. While an investigatory body later cleared Grissom of allegations that he had accidentally triggered the hatch himself, accusations of poor performance lingered.[191] Grissom, despite the loss of *Liberty Bell 7*, retained Slayton's respect and remained in the astronauts corps, throwing himself enthusiastically into engineering work on the Gemini spacecraft.[192] His work on Gemini was so well-received by Slayton that Grissom was assigned to command Gemini's first piloted flight in 1965, as well as what would have been the first Project Apollo flight in 1967.[193] Alan Shepard, despite earning Webb's anger for going over his head and appealing personally to President Kennedy for a second Mercury flight, maintained a position of great influence even while grounded, and later received a coveted Apollo command.[194] In 1969, Tom Stafford sent his Apollo 10 lunar module into a tumble when he accidentally threw a switch; his skillful recovery of the vehicle, though, was roundly praised; he went on to command a second Apollo flight.

Certain highly-regarded veteran test pilots might escape criticism for mistakes, but for most rookies, even minor failures in orbit were serious matters, with NASA offering few second chances to astronauts who were inattentive, or, like Apollo 9's Rusty Schweickart, merely prone to space sickness. Astronauts who expressed their displeasure with the program through practical joking, intemperate remarks to the press, or a seeming lack of enthusiasm for their work also invited censure. Such astronauts, in protest, often accelerated problematic behaviors or withdrew, causing colleagues to retreat from their defense, a death spiral that could end an astronaut's career in months. Scott Carpenter, in his Mercury flight, encountered various problems with the vehicle's systems and ignored repeated instructions to minimize thruster firings, a failure that resulted in him running short of fuel during reentry, when vehicle orientation was critical to assure that he entered the atmosphere at the right angle to ensure both a safe return and a splashdown near awaiting recovery vessels.[195] (Defying expectations, the supposedly taciturn Grissom is reputed to have wept in mission control during Carpenter's harrowing reentry, convinced he would burn up.[196]) Carpenter's landing, 250 miles from the designated recovery zone, was a performance even the normally reserved Slayton regarded as "sloppy."[197] Doctors examining Carpenter after the flight found him alert, but to Kraft, this fact proved only that the examining physicians couldn't tell a bad astronaut from a good one.[198] Carpenter, to Kraft, had panicked during reentry, and Kraft swore that Carpenter would never fly again.[199]

Other astronauts, according to Cunningham, disliked the fact that Carpenter "wouldn't compete" in the various sporting activities that the astronauts enjoyed for their exercise regimen, choosing to fence with an instructor instead of playing handball with his colleagues. Even in failure, though, Carpenter enjoyed the benefits of the silence astronauts accorded their colleagues: "[i]t was part of the code that astros did not ask each other point-blank about professional screw-ups," Cunningham notes, "but within the group it was gradually understood that he had made his last flight." While Carpenter continued to work with NASA through 1967 and made significant contributions to Apollo, injuries sustained in a motorcycle accident ended his flying career, and he spent much of the mid-1960s occupied with deep-sea diving research, "letting his hair grow long" and taking up "music," to Cunningham's embarrassment.[200]

Gordon Cooper, meanwhile, was given to provocative boasting, was unwilling to accept NASA authority, made indelicate remarks to the press, and acted out with jets and fast cars. The last of the Mercury astronauts to fly, Cooper quite nearly became the first astronaut dismissed by the agency. Increasingly frustrated by his antics, Mercury operations director Walt Williams and others at NASA Headquarters debated dropping Cooper from the Mercury roster; to Williams, former operations chief at Edwards, Cooper's bragging and stunts smacked of poor motivation. Slayton, though, came to Cooper's defense, confirming that he was competent to fly and suggesting that NASA dismiss Cooper immediately if it had no intention of flying him, an act that would have been a public relations disaster.[201] Further jeopardizing Cooper's career, though, the day before the scheduled lift-off in 1963, he buzzed Williams in an F-102 fighter jet, annoyed that engineers and flight surgeons had modified his spacesuit in a way he found uncomfortable, and which he feared would compromise its integrity in flight. Again, Slayton rose to Cooper's defense, narrowly averting Cooper's replacement with Shepard on the *Faith 7* mission.[202] Despite performing to NASA's satisfaction on flight (and even beating Schirra's pinpoint landing by half a mile), Cooper's flying career was abbreviated. He flew again in Project Gemini, but never again received a prime crew assignment. By 1969, Cooper, already criticized for his showboating and extracurricular auto racing, was earning poor marks from colleagues for his work backing up Apollo 10; he retired in 1970.[203] Flight Director Gene Kranz described Cooper as a "loner" and a "rebel," a characterization Cooper accepted (and Tom Wolfe greatly expanded), but the cause of Cooper's falling out with NASA is still subject to some debate.[204]

Like other twentieth-century workers, astronauts eventually encountered a system of control managed by senior colleagues. Having ascended to positions of authority in the space program, more junior astronauts chose to perpetuate these systems of control rather than undermine them, and to cast aside astronauts who failed to meet their expectations. One might think that astronauts would have attempted to organize against unreasonable expectations, capricious leadership, or the encroachments of NASA management on their private lives. Accustomed to endless ranking and stratified by seniority and service record, though, astronauts jealously guarded their own status and seldom bargained collectively.[205] When in crisis, astronauts turned to support networks formed in the military, confiding in one or two trusted individuals rather than in group structures. More interesting to the men than improving the lot of the "average" astronaut was attempting to figure which of them was the "anointed" one who, by virtue of his crisp command of workplace problems, effective communication, and flying skills, would rise above the rest. When certain members of their group—dreamers, playboys, scientists—defied conventions, others watched in silent, morbid curiosity at their fate.[206]

Chapter 3

Scientists in Space

As a civilian organization tasked with knowledge production, the National Aeronautics and Space Administration (NASA) was obligated to demonstrate scientific value in its human spaceflight program. Yet during its first decade, NASA's experimental vehicles made poor laboratories, and astronauts—test pilots who valued engineering data obtained through precise flying—found extraneous experimentation a distraction. In planning Project Mercury missions, NASA walked a fine line between engineering and science: flights intended to prove out basic components and systems were simultaneously packed with hundreds of minute experiments and opportunities for observation. American scientists of the 1960s were deeply divided over the value of piloted spaceflight, with many decrying it as wasteful, and one scientist anticipating Project Mercury to be "the most expensive funeral man has ever had."[1]

To proponents of human spaceflight, NASA Apollo program manager George Low later explained to one US senator, astronauts could accomplish more, and at lower costs, than unpiloted vehicles designed to replicate their diverse skills, a statement challenged in the press from the dawn of NASA's existence.[2] While early missions would fly only military pilots, NASA promised that, in time, American researchers would fill the skies, fulfilling NASA's mandate to produce scientific knowledge for the benefit of all. And just as many believed that human aviators were necessary to space exploration, many of NASA's leading scientists believed that only the presence of trained humans aboard space vehicles could ensure that first-class science actually got done.

The idea of a scientist in space was not a new one in 1965. Nineteenth-century conceptions of human spaceflight often conjured up crews of adventurers and inventors blasted into space within cannon shells or other fanciful devices.[3] By the twentieth century, a space crew comprised mostly of scientist-engineers was commonplace in books and film. Even on larger vessels, no more than two pilots was typical; the bulk of the crew usually comprised scientists, engineers, or technicians, including, frequently, an older professor of physics or chemistry identified as the ship's inventor or the mission's principal investigator.[4] Mid-century spaceflight fantasies conjured by Wernher von Braun, Willy Ley, and other rocket enthusiasts in the United States often focused upon the technical aspects of the project instead of the travails of pilots.[5] Von Braun's 1953 *Conquest of the Moon* did away with pilots completely, placing a scientist in charge of the vehicle and a team of engineers onboard to maintain the navigational computer.[6] In keeping with decades of conjecture, the very first men NASA's Space Task Group had recruited for the job in 1958 were professional explorers and researchers.

To those inclined to favor robotic exploration over human spaceflight, though, it was unclear why scientists were preferable to test pilots in the nose cones of rocketships.[7] "Space gadgets require only small amounts of power, can tolerate extremes of temperature, and, being expendable, can be sent on one-way missions," physicist Ralph Lapp wrote in the *New York Times*, quoting recent statements by space scientist and American satellite pioneer James Van Allen and *Science* editor Philip Abelson. "Man's brain, a wondrously fashioned three-pound computer, possesses unique capabilities, but the care and feeding of man in space must focus on the 50-fold larger mass of his body."[8] Indeed, the earliest space vehicles were poor locations for research and experimentation that barely accommodated their small crews and taxed them with time-consuming flying and engineering responsibilities. The weak thrust of early launch vehicles limited the space and weight available for research personnel and their instrumentation. Capsule layouts favored crew couches, instrument panels, flight control sticks, and the accoutrements of fighter jets rather than those of research laboratories.

Most within NASA at the time of the first Mercury flights assumed that, in time, professional scientists would join test pilots in space, though when and how was left to future study. The arrival of scientists in the astronaut corps in 1965 aggravated longstanding controversies within the space program over its connection with scientific research. At the same time, it challenged the professional identity of pilot-astronauts, who NASA had forced to open their spacecraft to men with vastly different skill sets and motivations. Where an earlier group of test pilots successfully surrounded themselves with the trappings of professionalism, NASA's new "Science Pilots" were unable achieve the status and respect enjoyed by their peers or an occupational niche that they could fully dominate.

Science in Human Spaceflight

The relationship between America's human spaceflight program and its scientific community had not always been fraught with acrimony. Inclined to maintain a clear separation between secret military space projects and civilian space exploration, the Eisenhower administration had encouraged NASA to pursue primarily scientific goals. Such efforts would both distract attention from America's secret and possibly illegal satellite reconnaissance program, and demonstrate American fidelity to internationalism, in contrast to the more insular Soviet space initiatives.[9] At least initially, some within the

Kennedy administration shared this view, with President's Science Advisory Committee chairman (and former director of MIT's Research Laboratory of Electronics) Jerome Wiesner recommending against the continuation of Project Mercury in favor of research with more scientific merit.[10]

President Kennedy's desire to mount a more aggressive challenge to Soviet achievements, though, combined with considerable media and Congressional pressure, eventually produced a more intensive human spaceflight program, one that drew more heavily on military resources and personnel and operated under a timetable less conducive to patient experimentation.[11] Ironically, it was in space science that the United States, in 1961, held a more impressive lead over the Soviets;[12] Project Apollo, while making more resources available for spaceflight, pushed aside these efforts in favor of the Moon Race. To the public, as Walter McDougall writes, the space program "was about science, sometimes spectacular science," but it was, "mostly about spy satellites, and comsats, and other orbital systems for military and commercial advantage."[13] To be sure, space exploration during the 1960s meant different things to different people: to some it was a research program enabled by high technology; to many others, though, it was an "extension" or even an incubator of military programs.[14] By the time of the 1969 Moon landing, NASA strategy documents listed scientific benefits of post-Apollo spaceflight in third place, after economic and national defense interests.[15] Guided by competing objectives, NASA's engineering and scientific communities, during the 1960s, frequently clashed over funding, hardware, mission goals and even personalities.[16]

Human spaceflight (in the form of Project Apollo) quickly became the most expensive of NASA's endeavors, and while pure space science research benefited greatly under NASA funding, engineering expenditures for human spaceflight consumed the majority of the appropriations. NASA characterized non-Apollo expenditures on the basis of whether or not they supported the lunar program, a classification scheme that reduced all other space and Earth science research to peripheral status. Scientists in NASA often found themselves with ample responsibilities but no authority to ensure that funds intended for them were not diverted to engineering operations. Often the result of these expenditures were vehicles that scientists neither needed nor wanted—piloted orbital cruisers—intended to satisfy the needs of the Moon Race and not of research programs, which favored smaller more numerous probes or large orbiting telescopes and experiment platforms.[17]

To many of its critics in the science community, Apollo was a "mission looking for a science" rather than "science looking for a mission."[18] For NASA associate administrator for Space Science, Homer Newell, though, a subordinate role to Apollo was superior to none at all.[19] Newell, a mathematician, had risen to prominence in the Naval Research Laboratory after World War II, flying captured V-2 rockets. He had played a key role in the formation of NASA in 1958, and found himself, upon joining the agency, struggling to maintain its integrity as a scientific institution.

Behind the strategic debates, fundamental cultural differences within NASA separated individualistic scientists from methodical engineers with whom they had to work closely on many programs. While engineers and scientists frequently swapped duties (and, in NASA's early years, shared the same occupational classification), differences in education, class, and outlook separated the two groups. Scientists with PhDs regarded themselves as elite intellectuals who, through the grant-writing process, had become comfortable selling themselves as high-status individuals. Engineers, with bachelor's or master's degrees, were more inclined to see themselves as team players: modest and punctual. To some NASA scientists, engineers were "second-class" technical workers who could build devices invented by others but not innovate on their own. To engineers, the better-educated but dreamy scientists were perpetually tardy "prima donnas" inclined to treat the engineer on their team as a "gofer" or to complicate schedules with capricious hardware changes.[20]

Astronauts identified with engineers in these disputes. Early in Project Mercury, astronauts had derided space scientists for cluttering flight plans with "Larry Lightbulb" experiments: nonessential apparatus and activities that used up valuable mass and volume allowances and distracted flight crews from engineering activities.[21] Rather than viewing themselves as scientists, astronauts self-identified as "technicians" comfortable with any form of instrumentation or machinery. Astronaut recruitment and training emphasized broad technical competence and maintained the fiction that they were technical generalists able to handle any flight regardless of its scientific objectives. "Deke's boys could handle just about anything that would interest the scientific lobby," Cunningham wrote, "and get the spacecraft back to earth as well. To fill a technical job, one sent the best technicians available."[22] In the scientific world, technicians were subservient to researchers and tasked with operating equipment under the direction of scientists; astronauts, though, clung to the job title as a mark distinction, in contrast to the scientists, who, the astronauts intimated, were eggheads and poor machine-tenders.

To astronauts, what made them effective was a common set of skills and traits. What made "scientists" valuable, though, was their uniqueness: each "scientist" might be an astrophysicist, geologist, physiologist, or any of a dozen other specialties relevant to the space program; and their chief credential, the PhD, had derived from original research. While eventually forced to add scientists to its astronaut ranks, NASA was not prepared to recast its idea of the space pilot; the astronaut corps would homogenize the scientists instead of fully harnessing their unique abilities.

Birth of the Scientist-Astronaut

It was in this unstable and often unfriendly climate that NASA's scientist-astronauts were "born." Increasingly dismayed by space science's diminished role in NASA, members of the National Academy of Science's (NAS) Space Science Board, chaired by Lloyd Berkner, met in the summer of 1962 to address the growing controversy over the role of science in human spaceflight and to suggest corrective measures. A physicist and leading figure in postwar American science policy, Berkner had made substantial contributions to the development of radar and to the study of Earth's upper atmosphere. He remained a leading space science authority through the 1960s, helping to guide NASA policy toward basic research.[23] NASA had, at first, sought to bulk up its scientific programs by establishing consulting relationships with scientists at sister agencies, but these were stopgap measures. The addition of scientists to NASA's flight roster was the easiest NAS recommendation for NASA to accept, as Homer Newell was of the opinion that it was unlikely that the agency would permit a scientist aboard the first lunar landing mission or establish a new institution to train scientist-astronauts—alternatives the NAS had considered.[24]

Since pilots already possessing advanced scientific training were uncommon, it was not clear whether pilots would be brought into the space program and trained as scientists or if scientists would be recruited and receive flight training thereafter. Slayton advocated the former, Homer Newell, in a speech before the AAAS in December 1962, suggested the latter. "I wasn't quite sure what I was supposed to do with them on flight crews," Slayton later recalled of scientists.[25] The NAS's 1963 *Review of Space Research*, though, produced from the deliberations of a 1962 summer study, elucidated three categories of scientist who might potentially fly in future space projects (in addition to a fourth group of "ground scientists"). These included pilot-astronauts with additional science training ("astronaut-observers") and scientists with some degree of flight experience: "scientist-astronauts" trained as full crew members and "scientist-passengers" drilled only in emergency procedures.[26] Project Apollo placed substantial responsibilities on all of its crew members, suggesting that only those with superior flight training—astronaut-observers and scientist-astronauts—would participate. For Slayton, flying a nonpilot scientist was out of the question; in his memoir, Slayton recounted an incident in which a navy flight surgeon at Houston's Manned Spacecraft Center lobbied for a Gemini assignment; Slayton rebuffed him, with Robert Gilruth's backing.[27] At least initially, NASA appeared enthusiastic only about recruiting physicians into the astronaut corps, and then, according to one NASA memorandum, only "to perform clinical procedures which would be dangerous when performed by laymen."[28]

As to the very real prospect that future "scientist-astronauts" might never actually fly in space, some at NASA were circumspect, but nonetheless continued their efforts to recruit them. Many within NASA expected the agency's scientist-astronaut program to produce only washouts and backup pilots, and were satisfied even with these limited aims. NASA director of Systems Studies, William Lee, hoped that the possibility of a lunar flight at some point in the future might induce more scientists to join NASA's ground operations and that, in any event, training "non-test-pilots" for flight work would be a useful experiment for the agency.[29] To some of NASA's managers, the fact that so few among the selected applicants would actually fly in Apollo was their most attractive feature, a fact also helping to smooth the feelings of others within NASA who were horrified by the thought of any scientist joining the astronaut corps at all.

NASA managers, though, felt that the agency would not be able to attract scientists to Houston to participate in a space program, given the piloting requirements of planned space vehicles and the "outspoken antagonism" of the "more verbal sections of the scientific community" to NASA's human spaceflight program.[30] Astronaut selections in 1962 and 1963 recruited pilot-astronauts with ever-increasing academic qualifications but were not successful in recruiting many with proper scientific credentials. Most of the Original Seven astronauts had received engineering or general science degrees, though two, Carpenter and Glenn, had college credits but had never received their bachelor's degree; all of the 1962 group pilot-astronauts were bachelor's degree holders, while three possessed master's degrees in engineering.

For the 1963 selection, NASA relaxed the requirement for specific test pilot training in the hope that doing so might attract more pilots with scientific experience, a move that was only modestly successful, attracting several more advanced degree holders and two astronauts, Walter Cunningham and Rusty Schweickart, with professional research experience ancillary to their piloting work.[31] Only Buzz Aldrin held a doctorate (from MIT), and it seemed unlikely that more scientists would enter the space program unless NASA radically altered its selection criteria. Aldrin, with an obsessive interest in gravitation and the mathematical principles of orbital rendezvous (and, according to colleagues, an inability to talk about anything else), was the subject of endless teasing among his peers.[32] In a 2007 talk, he forthrightly defied astronaut custom declaring himself a "spaceman" and not an aviator.[33] To Slayton, though, the credentials of some of his newest astronauts were no substitute for test pilot experience: NASA had simply "lowered" its standards to take Aldrin, Cunningham, and Schweickart, and Slayton responded by shunting the men to the bottom of the flight roster.[34]

As a forerunner to the problems that NASA's scientist-astronauts would later face, pilots of the 1963 group who showed the greatest scientific fluency received the most awkward reception. Cunningham, whose memoir provides the most vivid account of the astronaut corps in the mid-1960s, emphasized his own scientific expertise to distinguish himself from his aviator colleagues, but at the same time, clung furiously to his identity as a pilot. "From the beginning," Cunningham recounted, "I suspected that we [Rusty Schweickart and Cunningham] had been selected as sops to the scientific community, but it took...the National Academy of Sciences...only a little while to find out that we were just like the rest of the 'dumb fighter jocks.' "[35] Fighter jocks, but not true test pilots. Cunningham, according to fellow astronaut Gene Cernan, arrived at NASA labeled as more scientist than pilot, and, like Schweickart, found only a single flying opportunity in Apollo.[36]

Unsatisfied by the 1963 selection, proponents of the scientist-astronaut concept continued their efforts over the next two years, with NASA geologist Eugene Shoemaker, psychologist Robert Voas, and engineer-managers

Joseph Shea and George Low, debating details of the potential recruitment program, including the sciences to be represented and the degree of flight experience expected of candidates. Ultimately, NASA officials expressed a preference for geologists and physicians, the former of which might support lunar research and the latter, long-term spaceflight projects. The issue of whether the scientists would need to be pilots, though, was never adequately resolved; nonpilot scientists were obviously more numerous but more likely to wash out of training.[37] The eventual recruitment announcement of 1965 welcomed applications from physicians and scientists with or without flying experience but able to pass a military flight physical. Civilians selected for the program received NASA salaries at government scale, although the proper salary level for the men was to be a subject of debate within NASA for years afterward.

The 1965 Selection

NASA's eventual recruitment notice for scientist-astronauts read like an elaborate call for a space science graduate program. Applicants would need to submit "transcripts of academic records" and those without current scores would need to sit for the Graduate Record Examination, customarily administered to college upperclassmen. Subsequent requests for information, NASA indicated, would include a portfolio of scholarly work, information about extracurricular activities, and possibly a personal essay containing "individual thoughts on scientific objectives of manned space missions."[38] With scientist-astronauts expected to have piloting responsibilities, NASA had decided to give preferential treatment to applicants already with flight experience. Requirements that applicants be under 36 years old and possess 20/20 uncorrected vision further guaranteed that older scientists were unlikely to qualify, leaving only extremely fit junior researchers available for consideration.[39]

Applications were passed to the NAS for initial examination. The two-stage evaluation process conducted for the scientist-astronauts mirrored that of the pilot-astronauts, with the NAS vetting candidates' scientific credentials before proposing them to NASA for more intensive scrutiny. NASA's own Selection Board included, critically, its *de facto* chief astronauts, Slayton and Shepard, who participated in the panel interviews that closed the selection process. After three rounds of astronaut medical exams at the Lovelace Clinic, though, NASA abandoned the facility and its rigorous testing for a more limited evaluation program at the USAF Aeromedical Center in San Antonio, Texas. Instead of broad research, evaluators sought primarily to identify disqualifying medical defects like ulcer, diabetes, and gall stones. Stress testing at Wright Aerospace Medical Laboratory of the kind inflicted upon earlier test pilots was also abandoned (both for the scientists and a group of pilot-astronauts recruited the following year), marking the decline of the aviation medicine community's influence over astronaut selection.[40]

The 15 would-be scientist-astronauts who assembled in San Antonio, though, found themselves subjected to many of the stresses encountered by the original pilot-astronauts, and responded in much the same way, enduring examinations without complaint and competing with each other over physical performance, over and above what NASA required.[41] NASA clung to the hope of locating scientists with flying experience, and the 1965 selection attracted two individuals who were already jet-qualified, physician Joe Kerwin and physicist Curtis Michel, as well as several other individuals with flying experience and military backgrounds, including physician Duane Graveline and physicist Owen Garriott.[42]

Aeronautical engineer and newly minted physicist Edward Gibson heard about the selection from his wife, who read about it in the *Los Angeles Times*. "I thought she was making it up," he later recalled. While later scientist-astronaut selections drew more applicants without NASA, military, or piloting experience, Gibson was already a private pilot and had long dreamed of an air force career.[43] Of the men NASA ultimately selected, only geologist Harrison Schmitt lacked any piloting experience. All candidates, though, were enthusiastic for a ride in a NASA's T-38 jet trainer, a new addition to the examination routine that exposed nonpilots to their first taste of supersonic flight and high-g maneuvers.[44] Rather than fearing the flight, applicants seemed to have regarded it as an inducement. Echoing the folksy modesty of the first pilot-astronauts, Gibson even claimed to have submitted his NASA application merely for the free airplane ride.[45]

The selected scientist-astronaut candidates were also far fewer than NASA hoped for or expected. The 1965 selection produced nearly 1400 applications, but few applicants possessed the right combination of physical attributes and qualifications and NASA approved only six astronauts, half the number needed: Garriott, Gibson, Duane Graveline, Kerwin, Michel, and Schmitt (Figure 3.1).[46] To examiners, these quasi-military scientists were standouts in a troublesome applicant pool with a discouragingly large number of medical ailments. Compared to the test pilots the physicians examined in previous selections, the scientists were in poor health.[47]

For Gibson (and many young scientists to follow), selection as an astronaut candidate was a surprise that placed him in unfamiliar territory. A pilot in his late-20s or early-30s was likely to have amassed substantial work experience, but the scientist-astronauts of the same age had spent most of their lives in school. Having only recently occupied the lowest wrung of the scientific profession, the public now attributed to them knowledge, skills, and experience that they knew they lacked. Though each possessed a scientific specialty, the men also recognized that NASA regarded scientists as fungible commodities, and that geologic training would be most valuable, so all of the scientists boned up on the field, hoping to demonstrate their usefulness. Recognizing the importance of solar physics to subsequent Apollo missions, though, Gibson quickly shifted focus, reading up on solar physics to position himself on a future crew.[48]

For NASA, though, most of the scientists were, foremost, fundamentally deficient in one skill basic to all previous astronaut classes—flying experience—a deficiency NASA would remedy immediately by dispatching the

Figure 3.1 NASA's scientist-astronauts in 1966 from left: F. Curtis Michel, Owen Garriot, Harrison Schmitt, Edward Gibson, and Dr. Joseph Kerwin. Duane Graveline left NASA shortly after selection (NASA photo)

nonpilot members of the group to Williams Air Force Base near Tucson, Arizona for flight training. There, they underwent physical training, classroom instruction, and successive instruction in ever-faster and more sophisticated training aircraft alongside air force officers ten years their junior, many of whom were destined to fly in Vietnam.[49] Unable to discipline the four civilians, the air force, the scientists believed, was uncomfortable with their presence, but trained them nonetheless, and the astronauts emerged as above-average jet pilots. Kerwin and Michel were already qualified aviators and several others had some flight training. Only Schmitt was a true novice, and others, like Garriott and Gibson, looked forward to NASA's mandatory jet training and excelled at it.[50] The widespread conviction that scientists would make poor flyers was based more in stereotype than fact. Statistically, it was the pilot-astronauts, not the scientists, who were accident-prone, with 8 of the first 54 pilot-astronauts dying in flight mishaps. The scientists, by contrast, suffered no fatalities.[51] The delay created by their jet training, though, impacted every member of the 1965 group; Kerwin and Michel, already jet-rated, occupied themselves with support duties until their colleagues completed their flight training, not able to assume a position in the flight roster until their colleagues had caught up.

Not surprisingly, it was Kerwin and Michel who were the first to realize that flight experience might not make the men astronauts to the rest of the astronaut corps.[52] Kerwin understood the precariousness of his position at

NASA when he learned during a routine meeting that the agency would select a new crop of astronauts in 1966. When pilot-astronaut Dick Gordon asked if they would be professional aviators, Kerwin later recalled, Shepard exclaimed: “Well, I certainly hope so!”[53] “We were not regarded as really instrumental to what was going on,” Gibson recalled upon returning with his scientist colleagues to NASA after flight training. “I then realized that maybe they sent us off to flight school hoping we would quickly flunk out or kill ourselves, or, anyway, not show up back here.”[54]

NASA’s original, abortive 1958 call for astronaut “researchers” had anticipated inquiries from a variety of scientists and adventurers. NASA’s closed selection of military pilots set recruitment standards that pilot-astronauts resisted lowering to accommodate dilettante aviators.[55] In 1961, preliminary crew designations for Apollo floated the concept of an “engineer-scientist” flying alongside the pilot crew; by 1965, though, NASA’s pilot culture had grown so monolithic that the Astronaut Office could scarcely countenance the idea of anyone, who was not called a “pilot,” flying in space [56] Even as late as September 1967, astronaut meetings were still called “pilot meetings,” despite the arrival of the first scientist-astronauts two years earlier.[57] Early formulations of crew titles for Apollo had contemplated a variety of designations, including “commander pilot, navigator co-pilot, and engineer-scientist,” and, briefly, “command pilot, senior pilot, and pilot,” but as former jet fighter pilots, astronauts were accustomed to having sole responsibility for flying their craft and refused to be labeled as co- or junior- pilots.[58] In a November 29, 1966 memo, Deke Slayton settled the matter of job titles in Apollo flights, determining that the vehicle would carry two “pilots” in addition to the “commander,” each designated to fly either Apollo’s Command Module (CM) or Lunar Module (LM) as separate (and roughly equal) crew members. Whether all three pilots in Apollo would do any actual flying was a different matter altogether; even during lunar flight, the commander would actually land the LM, leaving the Lunar Module Pilot (LMP) with other responsibilities. On Apollo flights in which no LM was to be flown, Slayton ordered that the vestigial third crew member be called a “Mission Module Pilot,” apparently with the idea that scientist-astronauts would be piloting their experiments.[59]

While maintaining the fiction that all three crew members needed to be test pilots, though, Slayton conceded that the third crew member was less than critical and occasionally earmarked the seat for “weaker” astronauts still needing to prove themselves.[60] This seat was the only one scientist-astronauts would ever fill in Apollo. Chosen mainly for their scientific qualifications, none of the scientists could fly as Commander or Command Module Pilot (CMP), the most flying-intensive positions and two of the three seats available on any Apollo flight. The Astronaut Office, and Slayton in particular, considered spacecraft to be experimental vehicles not yet ready for “scientific” missions; flying amateur pilots on them, Cunningham recounted, might, in Slayton’s mind, lead to disaster:

> Deke maintained that the development of a highly experimental flight test program like the development of manned spacecraft required specialists who

> had dedicated their lives to the operation of the closest related kind of equipment. The program could not afford the luxury of anyone who couldn't carry his own weight in such a flight test program.[61]

In flight assignments, thus, scientist-astronauts always remained "at the bottom of the pecking order."[62]

Upon the arrival of scientist-astronauts, though, even the most junior pilot-astronauts suddenly had colleagues to whom they could feel superior. Publicly, NASA never conceded that any of its astronauts were less qualified than any others; even the scientist-astronauts had received jet training and were proficient aviators. Privately, though, an organization that distinguished military test pilots from military operational pilots found the scientists to be a luxury—and a liability. "We quickly decided that the new breed was inferior" wrote Cunningham.[63] The astronauts had spent their professional careers sizing up other pilots: as test pilots, part of their duties involved anticipating how design flaws might befoul inexperienced pilots, and determining how they might be corrected. This training encouraged a patronizing attitude toward less experienced aviators. Pilot-astronauts, whether in jest or in genuine disapproval, had always attacked pilot colleagues by poking fun at their flying skills. Jim Lovell endured the nickname "shaky,"[64] while others spread rumors that Alan Bean was afraid to fly at night.[65] Pilots who did not fly with the enthusiasm of their colleagues were openly mocked: when Scott Carpenter eagerly sought time in capsule simulators instead of jets, Slayton, Carpenter recalled bitterly, yelled at him to "fly the goddamn plane for a change."[66] Astronauts who teased each other by impugning each other's flying abilities singled out the new scientist-astronauts for this kind of hazing as a group,[67] comparing them unfavorably NASA's few women "secretaries."[68]

Behind the pilot's dismissals lay fear of their own obsolescence. With their very presence in the Astronaut Office, the scientist-astronauts undermined NASA's prevailing "pilot" culture; they reminded test pilots how tenuous their hold on the human spaceflight program actually was, and they threatened to take some of the last crew slots available in the Apollo program. More significantly, NASA's primary mandate to expand "human knowledge of phenomena in the atmosphere and space"[69] made scientist-astronauts the future of the human spaceflight program, and the pilots knew it. Noted Cunningham, "[s]cientist-astronauts were brought into the program as far back as 1965. It was clear even then that they would outnumber the aviators some time in the future. . . ."[70]

The arrival of scientists might have brought NASA more closely in line with its own charter, but it would mean an inevitable dilution of the pilot-astronauts' reputation and status. Among the many arguments that had been used to deflect efforts to enlarge the astronauts corps to civilians was the fear, expressed by George Low in response to questioning by representative James G. Fulton (R-PA) at a July 1962 hearing, that employing crew members other than highly experienced military test pilots would reflect poorly on the technical competence of an agency still struggling to establish its legitimacy.[71] The astronauts felt the same way, noted Cunningham, and "worried that Congress and the public might not know the difference," between pilot- and scientist-astronauts, "or even care." Even worse, the presence of "milquetoast academic types" in the program suggested that many more people could be astronauts than actually were.[72] Competent or not, the scientists threatened the pilot's status, and the pilot-astronauts decided, quite consciously, to undermine them.[73]

As the new face of American "space science," the new scientist-astronauts would need to become convincing public spokesmen for NASA's haphazard scientific efforts. Though few in number and well-prepared in comparison to later scientist-astronauts, this task proved extremely vexing for them. Kerwin, a physician, was surprised when Shepard suggested that he abandon clinical practice to devote himself more fully to astronaut training, advice that seemed to him to suggest that his medical proficiency would be secondary to his spaceflight duties. Success in NASA, Kerwin quickly determined, would be determined principally by the degree to which scientists "subordinated our scientific interests to the program."[74]

Having done, so, though, the scientists still found themselves perpetually last in an astronaut food chain that continued to grow every year. As an astronomer, Curt Michel expected to play an active role the lunar program. His efforts to maintain ties to his academic work at Rice University proved challenging given his NASA schedule, and he received little sympathy from Shepard, who generally dismissed non-NASA work as a distraction.[75] Astronaut work, Shepard explained, would quickly expand to fill all of a candidate's waking moments, making outside activities—even those essential to a scientist's professional identity—virtually impossible. While NASA publicly entertained the idea that the scientist-astronauts would fill Apollo's lunar crews, those within the Astronaut Office seemed to know even by 1965 that the earliest (and, eventually, only) Apollo flights had been already earmarked for pilot-astronauts. The 1965 group scientist-astronauts, Michel later noted, expected that, at the very least, all of their numbers would fly before any astronauts selected after them (a NASA tradition up to that point), and were shocked to find themselves passed over for pilot-astronauts selected in 1966.[76] Because scientist-astronauts required flight training after arriving at NASA, any seniority they claimed was an illusion. "Until scientist-astronauts finished flight school," Slayton declared in his memoir, "they were only candidates as far as I was concerned."[77]

1967 Selection: "The Excess Eleven"

Most astronauts of the early 1960s could hope to fly within five years of joining NASA, first as copilots of more senior or veteran colleagues and then as commanders of their own crews. By the end of 1966, NASA, at the peak of its funding, appeared to be confronting a crew shortage brought on by the success of Gemini and build-up to Apollo. Not everyone agreed; Slayton (who had told the 1963 pilot-astronauts that NASA could not guarantee them all a flight in space), feared an overexpansion in the astronaut ranks.[78] Only months earlier, NASA had selected 19 new pilot-astronauts, and while NASA remained hopeful that Apollo and its successors would continue indefinitely, Slayton, concerned that large number of rookies were already piling up, feared a decline of the program budget that would limit flying opportunities for the newcomers. While NASA continued to add astronaut personnel through the late 1960s, though, fewer of the newer astronauts would eventually fly in Apollo, and even fewer would obtain one of the two "landing" seats on any of the six successful lunar landing flights.

On September 26, 1966, NASA announced its second selection of scientist-astronauts. Expecting a high washout rate among the new selections, Slayton sought between 20 and 30 new men, none of whom Slayton expected would ever fly in space. Again, the science community expected that any scientists selected might fly very soon after their training had been completed; again, NASA concealed the precariousness of the scientists' situation. Two of the new pilot-astronauts from the 1966 group, Bruce McCandless and Don Lind, had substantial scientific backgrounds (including, in Lind's case, a PhD) that NASA effectively treated them as scientist-astronauts, denying them training opportunities that might enable their assignment to a Moon mission.[79] (Don Lind held the record for the longest wait between selection and first flight, at 19 years, until teacher-in-space Barbara Morgan unseated him, with an improbable wait of 21 years.)

To NASA's pilot-astronauts, the latest scientist-astronauts had arrived at the worst possible time. In January of 1967, Gus Grissom, Ed White, and Roger Chaffee were killed in a prelaunch fire that destroyed their Apollo Command Module and pushed the vehicle's maiden voyage into 1968. The astronauts had died as engineers, not as pilots, attempting to fix the capsule's balky communications gear, and had been consumed by noxious gases and flames when an arc of current from exposed wiring ignited an aggressive fire in the cabin's oxygen-rich environment. In the controversy and investigation that followed, astronauts did not, as one might expect of shop floor workers, rail against the NASA for the deaths. Doing so would have done them little good, and as backup and support crew members for the mission, many felt complicit in the fire. "We—...all of the astronauts working on Apollo—we were as guilty as anybody over that particular tragedy," one astronaut recalled, "we knew that the spacecraft in those days was not very good." Succumbing to "go-fever," the astronauts proceeded despite the obvious dangers, convinced that their piloting and engineering skills would compensate for any deficiencies of the vehicle. "We believed our own publicity, that we could fly the crates [the spacecraft was] shipped in."[80] Like other engineers, the astronauts recognized both their personal responsibility for the agency's failures and a connection with management that made them unwilling to, and incapable of, standing against it in an organized way.[81]

The accident seemed to reinforce Slayton's insistence upon the experimental nature of the Apollo vehicle, but the lesson consumed the lives of three highly valued pilot-astronauts and seriously disrupted the flight schedule. Astronauts again proved central in both investigating the accident and in reconfiguring the vehicle, and the incident further reminded many that Apollo was a test program many years from the point where nonessential crew members could fly. In order to maintain the goal of a lunar landing by the end of 1969, NASA eventually eliminated two planned test flights, further reducing opportunities for pilot-astronauts in a space program that already appeared to be contracting. Yet despite the bottleneck, NASA continued in its search for new scientist-astronauts.

Noticing an advertisement on a bulletin board while finishing his doctorate at Yale University, physicist Joe Allen later claimed to have written to NASA merely out of curiosity about the selection process. He received NASA's response days after the Apollo 1 fire, and though it had been sent weeks earlier, Allen was vaguely disturbed by the letter, which suggested that NASA was trolling for astronauts to replace those killed in the accident. Nevertheless, he and dozens of other applicants eventually submitted to NASA's medical evaluations (which struck him as excessive and intrusive) and interviews before Slayton and Shepard (among others) and a ride in a T-38.[82] Psychiatric evaluation, trimmed to only a few hours from the several days of Project Mercury, had revealed the scientists to be, in most cases, identical to the pilot-astronauts in their confidence, lack of introspection, emotional distance, and enthusiasm for mastery.[83] Rather than identifying born astronauts, the examinations seemed to screen for a new American psyche: poised, ambitious, and excellent at taking exams. "[T]ests" Allen recalled with pride, "deemed us to be crackerjack pilots, although none of us had ever flown before." Yet their lack of flying experience and rigorous motivation to master the engineering aspects of flight marked the new applicants, like the 1965 scientist selections, as unreliable junior astronauts requiring remedial flight training. If the selection process was meant to locate those applicants with the best chance of becoming space pilots, it did not succeed. Several washed out or abandoned the effort after a few years; ultimately, Allen recalled, "the wisest brains at NASA were unable to administer tests that could identify a good pilot from a bad [one]."[84]

So enamored of the pilots had the psychiatrists been in 1959 that they virtually ignored the often troubling responses pilots provided for wanting to fly in space, assuming the desire to be a natural extension of an aviator's professional calling. Despite being scientists themselves, though, the psychiatrists couldn't figure out why the would-be scientist-astronauts would want the same opportunity, and were even less certain which responses would be pathological.[85] Allen, considering astronaut selection while at the University of Washington, suspected that Apollo follow-on missions might offer opportunities for professional scientists like him. Yet upon his selection, Allen claimed almost complete ignorance of space program.[86] Astronomer Brian O'Leary recalled being asked his motivations only once during selection, and providing a response so short and ambiguous that even he seemed disappointed by it.[87] Eventually, the examiners stopped asking and hoped for the best, but some of the selected scientists would prove quite willing to leave the program if it displeased them, and even publicly excoriate it.

Despite receiving 923 applications, 69 of which met NAS-NRC's standards, NASA examiners found only 11 candidates sufficiently qualified, less than half the number desired. Many of the selected scientist-astronauts, unlike the often arrogant test pilots who preceded them, seemed "surprised" to learn of their eventual selection.[88] The group—three physicians, two physicists, two engineers, a chemist, and three astronomers—ranged in age from 27 to 38, with the average age approximately 31, only slightly younger when selected than the average pilot-astronauts and the previous scientist-astronaut group. Having endured more than a decade of postsecondary schooling, some also lacked the work experience and rough-and-tumble worldliness of NASA's pilots, and appeared to them as schoolboys fresh from class. None could fly, and only physicians Franklin "Story" Musgrave and William Thornton, and engineer (and Australian native) Philip Chapman had military experience. Chemist (and UK native) John Anthony "Tony" Llewellyn had received his doctorate in 1958, but others were civilians fresh from the academy.[89] Physician Donald Holmquest had just received his medical degree when NASA selected him, and received his PhD in physiology in the following year. Joe Allen and engineer William Lenoir had received their doctorates in 1965, geologist Anthony "Tony" England did not receive his PhD in geophysics from MIT, until 1970. Astronomer Karl Henize was an established scientist with the Smithsonian Institution and Northwestern University; Robert Parker taught at the University of Wisconsin, but Brian O'Leary, the youngest of the men selected, had just received his PhD from Berkeley.

Criticism of the scientists by pilots occasionally smacked of class-consciousness, with the scientists portrayed as overgrown graduate students less willing to endure NASA's quasi-military regimentation and discomfort and more apt to question established doctrine than previous astronauts. Like the pilot-astronauts, the scientist-astronauts were solidly middle-class, though unaccustomed to combat or the poverty and indignities of military life. While all of the Mercury astronauts had college credits, and certain of them, like Shepard and Schirra, were glib, polished sons of career military officers, others like Grissom and Slayton were "small-town boys"—taciturn, gruff, and given to profanity.[90] To these seasoned veterans, the new MDs and PhDs were soft, weak, straight out of "college," and unaccustomed to making the kind of rapid life-or-death decisions that might save or kill a crew. Principally, though, the scientists were—to the "aviator fellows"—incompetent in the cockpit.[91]

Junior even by scientist-astronaut standards and virtually without the military pedigrees and flying experience of their scientist-astronaut colleagues, the new selections found themselves at the bottom of a pecking order so stratified that, according to Cunningham (who possessed a keen eye for NASA's many status distinctions), even the scientist-astronauts of 1965 regarded them as inferior underlings.[92] Soon after the 11 scientists arrived in Houston, Slayton asked to meet them at the motel at which they were staying to break the news that NASA had no room in the flight rotation for more astronauts, seemingly wiping his hands of the new recruits. "[W]e have been told by the government to take you," Allen recalled Slayton telling him and his colleagues, "but we don't have a job for you, not any of you."[93]

Unbeknownst to the scientists, by December 1966, Slayton had already selected the nine three-man teams which would crew all of the upcoming Apollo flights, as he had done for Gemini years earlier.[94] He then offered the assembled scientists an opportunity to resign, assuring them that they would "make no enemies" if they did. Half-heartedly, Allen recalls, Slayton continued, promising the men ground assignments in NASA if they stayed in the Astronaut Office, but urging them not to "fool" themselves into thinking that they would "be space flyers." The group quickly dubbed themselves the "XS-11" ("Excess Eleven") mocking NASA's indifference to their arrival.[95]

For the members of the XS-11, their astronaut careers began with the same five months of classroom and field instruction the pilot-astronauts had endured; like the pilots, they found much of the classroom instruction in engineering aspects of spaceflight uninviting, but, as junior academics, also found them pedagogically unsophisticated. "They were unintelligible," Allen noted, "I was a teacher, so I could tell that."[96] After completing ground instruction, NASA dispatched the XS-11 to air force bases for jet training. Unlike the earlier group, though, the members of the 1967 selections were divided among several installations. Singly or in pairs, the scientist-astronaut candidates often found themselves the lone civilians among 40 pilot cadets, and both the air force and the scientists themselves felt uncomfortable with the arrangement.

Some of the scientists immediately took to the training and excelled; Allen earned first-place marks in his flight training, a trophy, and praise from Shepard, which he regarded with secret pride.[97] When asked in 2003 how useful flight training had been his work in NASA, though, Allen was ambivalent: "[p]robably not at all," he concluded, though conceding that it had sensitized him to the difficulties involved in working with complex engineering systems. Even Garriott, who professed no enmity for NASA's pilot-astronauts, felt that the scientists were sent to flight school not to gain valuable skills but to "prove" themselves in the kind of high-stress work that pilot-astronauts undertook routinely.[98]

Some of the scientist-astronauts, though, refused to submit to indoctrination. For NASA's pilot-astronauts, military discipline was routine, but for many of the new scientist-astronauts, it was an irritation, and one for which they showed disrespect and occasional open contempt. Chapman, O'Leary, and Parker earned their flight instructors' ire by either refusing (in breach of military discipline) to wear caps while on the ground,[99] or refusing to wear matching ones.[100] Chapman recounted being harassed by colonels for the infraction, at least until they determined that he was a civilian and beyond their authority, prompting an apology. Eventually, Chapman recalled, he made a point of making direct eye contact with senior officers, who were puzzled by his confidence, and, especially, the lack of rank insignia on his flight suit. Chapman (like O'Leary) found the military aspects of NASA training absurd, and delighted in mocking his air force hosts, coaxing, on one occasion, a salute from a major general concerned that someone as audaciously underdressed as Chapman was likely worthy of deference.[101]

As Slayton had promised, though, completion of 53 weeks of flight training did not lead to crew assignments for the scientist-astronauts. Among NASA's pilot-astronauts, Slayton enforced a policy that no member of a later-selected group would fly before all of the members of the previous group had done so, but for scientist-astronauts, the rule did not apply. At the same time that veteran astronauts jockeyed to command Apollo lunar flights (and junior pilots hoped to beat the odds and join them), scientist-astronauts found themselves in a parallel world of training and support work, with flying an increasingly distant dream. With ground and flight instruction complete, the scientist-astronauts were, like their pilot colleagues, immediately assigned to assist teams preparing hardware and crews for upcoming Apollo flights. Where possible, NASA attempted to assign astronauts to their areas of expertise, but with geology dominating the lunar missions, some scientist-astronauts found themselves working outside their individual areas of expertise in an organization that often seemed to regard scientists as fungible commodities.

Principally, Joseph Allen later noted, NASA called upon the scientist-astronauts to translate the needs and interests of pilots and ground scientists, placing them in the middle of the often intense mission politics in Houston. The scientist-astronauts could communicate between camps, but were powerless to mediate the relationship between the competing agendas of flight directors and NASA's considerable army of ground scientists—the former who wanted astronauts off the Moon as soon as possible after arriving there, and the latter who were intent on squeezing from Apollo flights as much data as could be obtained.[102]

By 1969 (when the second group of scientists finished flight training), NASA's employment rolls everywhere but at the Astronaut Office had dropped by a third from their 1965–66 peak, and human spaceflight appeared threatened under President Richard Nixon.[103] The first members of the 1966 pilot-astronaut selection—Fred Haise and Jim Swigert—found slots in a prime crew with Apollo 13 in 1970. Even the imminent conclusion of the Apollo lunar landing missions, critically, did not end the reign of astronauts of the 1962 and 1963 groups; while many veteran astronauts left NASA at this point, several of those who remained—Pete Conrad, Tom Stafford, Alan Bean—commanded considerable influence in the program as well as in the command of Apollo crews of the postlunar period.

For most other rookies, NASA could offer only support crew positions that, unlike the backup crew slots, offered plenty of opportunities to attend meetings but no possibility of flying.[104] Scientist-astronauts, arriving in a program to support an extended lunar program that was now unlikely, were immediately shunted to the "support" role. Merely with their presence, though, the scientists often made the pilots nervous, and as representatives of the science community, seemed to epitomize all of the extraneous concerns that many pilots believed were slowing the space program down. While pilot-astronauts jealously guarded their control over the operational aspects of spaceflight and bemoaned their lack of influence over program management, 1967 scientist-astronauts like Chapman and Holmquest saw their colleagues as a powerless and superfluous labor force distrusted even by other astronauts.[105]

While NASA's scientific leadership and the NAS pressured the Astronaut Office to fly its scientist-astronauts (or "Science Pilots") aboard vehicles in the 1970s, Slayton opposed flying scientists on "hazardous" missions, a category into which, by his definition, almost all Apollo flights ultimately fell.[106] Every member of an Apollo crew played some role in controlling the vehicle, and with more experienced aviators always available and scientific objectives always secondary, scientist-astronauts found NASA unsympathetic to arguments that their scientific credentials outweighed their lack of test pilot experience. Even during ground tests, NASA hesitated to place its scientist-astronauts in charge of even more junior pilot-astronauts, assigning only one, Kerwin, to command a 1968 vacuum test of the Block I Apollo Command Module.[107] Despite Cunningham's concerns, NASA's efforts to integrate the scientist-astronauts in the projects of the late 1960s largely began and ended with their selection. The scientists were neither accepted as pilots nor valued as scientists, were forced to gather skills they would never use, and were prevented from contributing in their own disciplines.

The space program seemed to have so little idea about what to do with scientist-astronauts that examiners asked them in interviews, how much time the scientists expected to spend on research and what kind of research they would do if they found themselves in space.[108] The Manhattan Project, too, had required a large engineering endeavor, but scientists drafted to participate had arrived knowing what physical problems required immediate resolution. Unlike with other "big science" projects of the past, scientist-astronauts were not recruited to solve particular scientific problems; rather, NASA seemed to have selected them in the hope that the scientists would bring a set of research questions with them. Once selected, though, scientist-astronauts only infrequently found opportunities to engage in research, and spent the bulk of their time in vehicle training or supporting the work of ground scientists. Science consumed a quarter or less of the scientist-astronauts' time, and overt pressure from Shepard discouraged outside academic work. Even a scientist-astronaut, Shepard once chastised O'Leary, was an astronaut "twenty-four hours a day."[109]

At the same time, it tacitly denigrated their flying ability, the Astronaut Office neither cultivated scientific research nor provided the scientists with the kind of structural and institutional support they needed to generate scientific product.[110] Time, in particular, was lacking. Scientist-astronauts, instead, endured the same laborious and time-consuming training as their pilot-astronaut colleagues, including weeks of engineering briefings unrelated to their scientific specialties. To academic scientists accustomed to a "moral economy" emphasizing respect for scientific pursuits, mutual support for each other's endeavors, and an understanding of the complex demands upon a scientist's life, the Astronaut Office was an elementary school playground.[111] Older astronauts bullied younger ones, astronauts competed viciously for status and credit, and almost no one appeared to have any inclination to support NASA's scientific objectives.[112]

Frustration soon thinned the scientist-astronaut ranks; Brian O'Leary and Tony Llewellyn resigned in 1968 during jet training, finding it more difficult than they had imagined. When Slayton calmly asked O'Leary over the telephone why he wished to resign, O'Leary had responded, flustered, that flying wasn't his "cup of tea." Accounts differ as to whether Slayton was furious at the news of O'Leary's resignation or ecstatic; in either case, Slayton contained his rage and did not try very hard to change O'Leary's mind on the matter, agreeing that a space career was a poor choice for a timid flyer. To O'Leary's chagrin, though, Slayton later repeated his private comments to the press, an act that appeared calculated to embarrass him.[113] O'Leary's very honest, off-the-cuff remarks seemed to Slayton to distill and confirm every criticism previously voiced about the scientists—that they were uncommitted, weak, and were poor aviators. Even the manner in which they expressed themselves bore this out; O'Leary's verbal insouciance stood in stark contrast to the speech patterns of Slayton, a man so given to halting grammar and expletives that "people who knew him cringed every time he got near a microphone."[114] O'Leary's spontaneous answer appeared to confirm the opinion of gruff, tough, and often profane Slayton that civilian scientists lacked the fortitude for spaceflight.

In an opinion piece appearing in the *New York Times* in 1970 and in his subsequent memoir, *The Making of an Ex-Astronaut* (1970), O'Leary responded, offering the unexpressed complaints of many of his colleagues at the time. One eventual complaint of scientists during Apollo was that engineers scheduled flights too frequently for adequate analysis to be undertaken of the data that each mission yielded. Not only were the Apollo flights "one technical stunt after another, with only minor increments in scientific return," but scientist-astronauts couldn't even get in on them. "While the scientist-astronauts are waiting a decade or more for a space flight," O'Leary fumed, "only test pilots are being flown." O'Leary's 1965 scientist-astronaut colleagues, "high performance jet pilots with years of astronaut training," had waited patiently while pilot-astronauts selected after them flew. Referring to the explosion aboard Apollo 13 months earlier, O'Leary insisted that scientist-astronauts could have handled the crisis as well as the all-pilot crew.[115] If the test pilots looked at space mishaps, and wondered how the scientists could do any better, O'Leary looked to the pilots and their frequent in-flight mishaps, and wondered how scientists could do any worse.

O'Leary's memoir expounded on these criticisms, and ripped into almost all of NASA's astronaut selection, training, and flight assignment protocols. With all of the active-duty astronauts bound by press contracts controlling their life stories, none could afford to be as brutally frank in their observations. O'Leary's criticisms of NASA began with his selection and ended with his resignation. NASA was steeped in a quasi-military piloting culture that denigrated science; to a scientist-astronaut, nothing about it appeared to make any sense. Though none of the Excess Eleven would be allowed to pilot a space vehicle, for example, all expended substantial time learning to fly supersonic jets and maintaining their proficiency. O'Leary in particular recoiled at the danger and wondered why all the fuss was necessary. Among the answers Slayton provided was that Soviet nonpilot cosmonauts tended to get sick in space, a comment that seemed to label scientists as weaklings.[116]

As the scientists languished on the ground, the original ideal of trained astronaut-observers picking up the scientific slack in orbit never materialized. The scientific product the pilot-astronauts produced, O'Leary concluded, was shoddy; they mishandled equipment, took useless photographs, and failed to make adequate records of their findings.[117] The pilot-astronauts, O'Leary complained, proved willing neither to behave as scientists themselves nor to accept scientist-astronauts as equal partners able to make important contributions to the human spaceflight program. When NASA administrator James Webb met with the astronauts in 1968 to break the news about an expected human spaceflight hiatus during the early 1970s, Chapman questioned him about the impact upon the scientific community. "To hell with the scientific community," boomed influential pilot-astronaut Frank Borman, upon which other astronauts began laughing. The scientists were demoralized; Borman's comment was "like an electric shock."[118]

To Slayton, though, complaining scientist-astronauts were merely weak and unmotivated. O'Leary, nearsighted and partially colorblind, was, recalls Slayton in his memoir, a "young guy. . .straight out of college with a lot of mistaken ideas," who had never completed flight training. To Slayton, O'Leary's difficulties in flying were indicative of youth and of personality traits incompatible with spaceflight. Rather than rally to O'Leary's side, though, fellow scientists-astronauts abandoned him, surprised by O'Leary's attitude toward flying but not surprised that the unenthusiastic O'Leary had chosen to resign. David Shayler quoted one unnamed scientist-astronaut as calling O'Leary "immature and unrealistic," speculating that the "egotistical" O'Leary had misled examiners during selection as to his motivations and willingness to be a "team player." O'Leary, while admitting to being young and somewhat arrogant, did not seem to be too much different from his equally confident colleagues, but unlike them, he refused to submerge his identity in a space program that, to him, seemed riddled with absurdities.[119] Such criticism, to seasoned astronauts, smacked of disloyalty: if new astronauts didn't agree with the old ways, it was because they didn't understand them, or wouldn't. To Slayton, O'Leary's public criticisms of the Astronaut Office only "confirmed" Slayton's "impressions" that O'Leary failed not because he had been mistreated, but because he "never was an astronaut" to begin with.[120]

Michel and Chapman complained about flight opportunities for scientists and NASA's lack of commitment to science, and eventually resigned as well.[121]

Led to believe, according to Michel, that some would fly as early as 1968, the 1965 group scientist-astronauts waited through the year as crew assignments went to other astronauts and NASA continued to add to its scientist roster. As the Apollo 11 lunar landing approached, only one of the scientist-astronauts of 1965 (and none from the 1967 group) was allowed to participate in the helicopter training considered necessary for astronauts undertaking lunar missions, an opportunity extended to the three-quarters of the 19 pilot-astronauts NASA selected in 1966. When Michel protested, Shepard yielded, assigning two more scientists to helicopter training (including Garriott), but snubbing Michel. To Michel, Shepard seemed to be simultaneously endorsing his recommendation and "rebuking" him for suggesting it.[122]

Undervalued as a pilot, Michel felt similarly snubbed as a scientist. Scientific instrumentation reduced the astronauts to semiskilled technicians, but the pilots preferred it that way; scientific duties were an annoyance better dispatched by machine. When the Bendix Corporation presented NASA with designs for a self-contained scientific package to use on the lunar surface, pilot-astronaut Bill Anders, objected, insisting that the Apollo Lunar Surface Experiments Package be reconfigured with the equivalent of a "Big Red Button" that would deploy the device with minimal astronaut involvement.[123] For the scientists, though, operating such instruments required little of the specialized knowledge they had gathered in their scientific careers. (Nor, as Chapman discovered, was the Astronaut Office particularly keen on the scientists designing their own in-flight experiments.)

Disheartened, and with no chance of flying in the foreseeable future, in 1968 Michel requested that NASA put him on leave so that he could return to Rice to resume his teaching and research for the academic year 1968–69, a request to which NASA grudgingly assented. When he returned, Michel discussed his future with Shepard and Slayton, the latter, griping about the recent growth in the Astronaut Corps, in fact seemed pleased to be reducing his headcount, regardless of the reason.[124] When Shepard and Slayton confirmed that only Michel's colleague Schmitt had a chance of reaching the Moon, Michel promptly resigned. An article in the *New York Times* concerning Michel's subsequent departure in August 1969 was entitled "Astronaut Resigns to Pursue Science," a telling indictment of NASA's human spaceflight efforts.[125] To Michel, "one ride in space" meant sacrificing a career of scientific work, and he was unwilling to make that trade-off.[126] The departure, like O'Leary's was somewhat bitter; Michel later claimed to have been snubbed by NASA, and criticized Cunningham's depiction of him as a "cynical and casual" iconoclast who NASA dismissed "as soon as it was convenient."[127]

In May 1971, MSC announced that Donald Holmquest would take a one-year leave to teach at Baylor University College of Medicine, ensuring his nonparticipation in the remaining Apollo flights.[128] Writing to NASA administrator James Fletcher in 1973, Holmquest later described a pattern of "discriminatory abuse" from the "ignorant and corrupt" Slayton from the moment the scientist-astronauts joined NASA, reflected in a "hostile" work environment and the scientists' "professional isolation." Too often, Holmquest continued, "NASA engineers and support scientists often pursue experiments because they are feasible and not because they have any scientific merit."[129]

To Chapman, as well, time spent in NASA was "out of the mainstream of science."[130] Occasionally, the interaction between scientist- and pilot-astronauts produced productive collaborations, as when pilot-astronauts cooperated enthusiastically with the scientists in organizing small experiments during their free time in space. At other times, Chapman recounted, Slayton obstructed their efforts. The earliest Apollo lunar landing missions had offered relatively little time for more than the most rudimentary scientific work, while on later flights, the disposition of the flight crew toward experimental objectives often had a profound effect on how much science would be conducted during the flight.[131]

Coordinating scientific work for Commander David Scott's 1971 Apollo 15 mission, Allen was delighted to find an enthusiastic crew eager to undertake serious geological study of the Moon. As a member of the support crew for the flight, Allen was even able to enlist Scott's participation in the recreation of Galileo's famous "Leaning Tower of Pisa" thought experiment, arranging for Scott to smuggle a falcon feather to the Moon and to drop it on camera, alongside a rock hammer, to demonstrate that, in the vacuum of space, the two objects struck the ground at the same time.[132] Though little doubt existed as to the outcome of the experiment, it proved extremely popular with educational audiences and the public.

When Chapman had earlier attempted to instruct enthusiastic pilot-astronaut Stuart Roosa where he might point his camera to make a scientific discovery while whiling away his time in lunar orbit on Apollo 14, Slayton, Chapman recounted, berated them both, threatening to bump Roosa from the flight and ground him permanently if, while in lunar orbit, Roosa looked at anything except what Slayton told him to. Slayton's principal concern was public relations—by limiting the number of experiments to a small number of approved activities, Slayton reputedly wrote in one memo, NASA could more easily claim 100 percent success in achieving mission goals. Fearing that Apollo would end before he received a crew assignment, Chapman joked with colleagues about contaminating lunar soil with modified goat urine in an effort to stir up funding for future missions.[133]

To Chapman, geologist Harrison "Jack" Schmitt was an obvious choice for Lunar Module Pilot on a lunar landing mission; in 1970, Schmitt was assigned to backup Dick Gordon's Apollo 15 crew, an assignment that, under NASA's rotation system, would place Schmitt in the prime crew of Apollo 18. In September 1970, though, NASA, under increasing budget pressure, cancelled two lunar missions and reshuffled the remaining flights, ending the Apollo program with mission 18 but renaming it Apollo 17 and placing it under the command of pilot-astronaut Gene Cernan. Gordon, incensed, lobbied with NASA to replace Cernan's crew with his own.[134] At the same time, though, NASA's chief scientist, Homer Newell, and the National Academy of Sciences were placing increasing pressure on Slayton to fly Schmitt as LMP on Apollo 17.[135]

Himself sharply critical of NASA's unwillingness to include scientists in flight crews, Newell was, in 1971, nonetheless challenged to defend the agency against critics curious why, after four attempted Apollo Moon landings (three of them successful) and two more missions already crewed, a geologist had not yet joined the flight roster. Newell described the Apollo 15 astronauts' extensive geological training and his intention that Schmitt would fly on Apollo 17, but his comments smacked of hope rather than certainty.[136] Slayton resisted flying Schmitt, reportedly noting that a "dead geologist is no use to anyone."[137] Even worse, were Schmitt assigned to Apollo 17, the crew member whose place Schmitt would take was highly respected pilot-astronaut Joe Engle, who so closely matched the ideal of the test pilot that he made other test pilots uncomfortable.

Joe Engle "had some flaws as an astronaut," later wrote Slayton, but "was a terrific stick and rudder guy."[138] He was a "protégé" of legendary aviator Chuck Yeager; a "natural" pilot who had flown the X-15 to the edge of space and beyond, 16 times even before joining the NASA Astronaut Corps. Cernan was uncomfortable having to choose between the two men, but he clearly preferred Engle over Schmitt. If Engle had a flaw, it was that he was "too good"; Cernan feared that he would skimp on academic and simulator training, but Engle still appeared a safer choice than Schmitt, whom Cernan did not even acknowledge as a "pilot."[139] While respected for his hard work in supporting other Apollo missions,[140] Schmitt fell behind even his scientist-peers in jet pilot training, and became a punch line. "If God had meant man to fly," other astronauts joked, "it wouldn't have been Jack Schmitt."[141]

Just before the launch of Apollo 15 in July 1971, Slayton secretly filled the last of the Apollo crews, excluding scientist-astronauts from the roster.[142] Scientist-astronaut Phillip Chapman was one of many individuals who visited James Fletcher to lobby for a scientist aboard the final Apollo lunar flight, the only argument with Slayton that Chapman claimed to have ever won. In August, Slayton and Cernan yielded, and Schmitt became the only scientist-astronaut assigned to the prime crew of a lunar landing mission (Figure 3.2). Dick Gordon, after a year in an administrative post, resigned from NASA. When Slayton finally announced, in the summer of 1972, that no more scientists would be assigned to Apollo flights, Chapman quit as well.[143]

Discussions over the goals of the new Apollo 17 mission epitomized the conflicts between NASA's space science community, which hoped for an ambitious final flight, and NASA's astronauts and engineers, who wished to minimize risk. Schmitt had hoped to exploit the final Apollo flight for a landing on an unexplored area of the Moon: the far side, which is never visible from Earth's surface. Such a landing would have severed the normal radio links between the crew and Earth and would have required a communications satellite to be placed in lunar orbit. NASA physicist and engineer Robert Farquhar, the Goddard Spaceflight Center scientist working on the proposal, recalled Jim McDivitt joking at one September 1971 meeting about a longtime conspiracy between Farquhar and Schmitt to pitch a far side landing. Schmitt, though, avoided confrontation in dealing with the

Figure 3.2 Commander Eugene Cernan and Lunar Module Pilot Harrison "Jack" Schmitt prepare the Lunar Roving Vehicle (LRV) as the Lunar Module receives a preflight checkout in preparation for Apollo 17's 1972 flight (NASA photo)

powerful McDivitt, utilizing the same interpersonal skills that pilot-astronauts deployed. Advising Farquhar on his presentation, Schmitt encouraged the innovative and outspoken engineer not to present his most ambitious and complex plans, fearing that pilot-astronaut McDivitt would not understand. "Jack Schmitt cried out, 'Oh my God, don't show that slide'.... Schmitt was concerned that Jim McDivitt would view this type of transfer as too complicated and too risky."[144] Though feasible, NASA chose not to pursue the far side landing; the December 1972 Apollo 17 flight was a success nonetheless, exceeding previous flights in the length of the lunar stay and in its scientific yield. Most importantly for the scientist-astronauts, the mission had proven that a professional geologist could perform in the astronaut role.

Laboratory in Space: Skylab

Upon completing jet training, the first class of scientist-astronauts was assigned to support work on what, by 1965, NASA had dubbed the Apollo Applications Program (AAP): a series of missions planned for the 1970s intended to mount extended Earth-orbit and lunar voyages. Upon completion of their flight training, the Excess Eleven joined them. Throughout the late 1960s, engineers had planned a variety of missions that would supplement Apollo vehicles with orbiting experiment modules and other hardware. With AAP, NASA hoped that astronauts of the 1970s would enjoy long-duration stays in Earth orbit and, eventually, on the Moon, creating a traditional laboratory work environment in space that would justify continued human spaceflight. Budget cuts eliminated most of these projects, but retained the space station concept—the spent upper stage of a Saturn IB launch vehicle, drained of residual propellant and pumped full of oxygen. Surplus Saturn V launch vehicles from cancelled lunar landing missions provided NASA with a rocket large enough to launch a fully fitted space laboratory—"Skylab"—into Earth orbit; Apollo vehicles would ferry crews to the station.[145]

As a program intended to make good on the scientific possibilities of human spaceflight, AAP was embraced warmly by the scientist-astronauts, but as Skylab's launch approached, the traditional controversies arose in crewing. On paper, Skylab would tax the flying skill of only the crew member charged with piloting the Apollo spacecraft to the station; scientific and medical duties would consume the others' time. Slayton, concerned about the engineering aspects of the project, included only a single scientist-astronaut aboard each flight. The Space Science Board and NASA's scientists, especially Homer Newell, argued furiously for the inclusion of second scientist on Skylab crews.[146] Other NASA scientists believed as late as 1970 that all three members of Skylab crews would be scientist-astronauts, as these men had been trained as "Science Pilots" and were technically qualified to fly.[147] Further complicating the participation of scientist-astronauts in Skylab were the ongoing demands of Apollo: as Slayton later recalled, he assigned several scientist-astronauts to support scientific objectives on later Apollo lunar landings, but these assignments "basically took these guys out of the running for the first (and as it turns out only) Apollo Applications crews."[148] Scientist-astronaut Joe Allen later confirmed this, noting that rookie astronauts supporting late Apollo flights could play little role in Skylab, moving to follow-on programs that would not materialize for several years.[149]

Slayton and Robert Gilruth eventually prevailed, limiting the scientists to a single seat on each three-man Skylab crew and forcing Newell to defend publicly a more limited role for the scientist-astronauts than he himself had wanted.[150] "Over sixty percent of crew activities are devoted to onboard systems as opposed to experiment monitoring or conduct," Newell explained in 1971 to a Brigham Young University geology professor. "And of the time devoted to experiments, less than twenty-four hours is still free or unplanned time. The planned time is devoted to work which does not require a scientist—but rather a technician. That twenty-four hours free time may well further decrease or even disappear prior to launch dates of Skylab missions." NASA simply had no need for scientists; three-fifths of the Skylab astronauts' time was committed to engineering activities, and the rest could be performed by relatively unskilled people.[151] To the scientist-astronauts, science appeared to be an afterthought even in Skylab.

When finalized, Skylab offered crew slots only for the three scientist-astronauts of 1965 who had not already flown or resigned, and none for members of the 1967 selection, whom NASA had selected specifically to staff the Apollo Applications Program. Skylab, Curtis Michel feared, existed only to justify continued human spaceflight, and used the promise of scientific research only as an excuse. Indeed, the station's automated systems did not appear to require the mental faculties or experience of a principle investigator. Skylab, for example, would mount a solar photo-telescope originally designed for an unpiloted probe; crew members would do little more than flip switches pointing the telescope and activating its cameras, functions that were originally intended to be performed by an automatic switch. Rather than performing actual science, Michel surmised, NASA had de-automated a robotic vehicle just to add a piloted requirement.[152] To ensure mission success, NASA cross-trained the astronauts in operation of the telescope and other critical activities, further ensuring that scientist-astronauts' unique talents would be relatively unimportant to the program.[153] Oddly, pilots eventually filled even the roles earmarked for physicians: to prepare for medical emergencies in space, prospective Skylab crews underwent medical training, including extracting teeth from patients at the Air Force Hospital in San Antonio, Texas.[154]

Why Did They Stay?

Not all of the scientist-astronauts condemned the treatment they received from NASA; Garriott insisted years later that "there was relatively little friction between the scientist-astronauts and. . .Slayton or Shepard," and "nearly none with other pilot astronauts," except for a few colleagues dismayed over "crew selection."[155] Crew selection, though, was the single most important aspect of the astronauts' working lives, and the accounts of Slayton and others reveal the test pilots' a deep distrust of the scientists. Even the widely respected 1965 physician-astronaut Joe Kerwin, who flew on the first piloted Skylab space station mission in 1973, occasionally chaffed at the awkward position in which NASA had placed the group. To Allen, pilot-astronauts could be welcoming "on an individual level" but they regarded his entire group with suspicion and concern.[156]

Some scientist-astronauts clearly regarded confrontation as ineffective and chose to accept the tacit policies of the Astronaut Office. Garriott's competence and refusal to complain eventually brought him more managerial responsibilities. Story Musgrave, O'Leary remarked, was the member of the 1967 scientist-astronaut selection who best understood the cultural environment of the Astronaut Office and proved most adept at satisfying its pilot constituency.[157] "He doesn't argue about anything," pilot-astronaut Donald Peterson later recalled of Musgrave. "Of course, we picked on Story all the time, but I think he kind of liked it."[158] Musgrave most aggressively pursued flight training. Flying with him, Ed Gibson recalled, was like flying with a "computer." "He was such a good pilot," Gibson recalled, "[h]e did almost everything that well."[159] Had NASA extended the Apollo program, the indefatigable and multitalented Musgrave would have likely been the first member of his class to fly in it.[160]

For most of the scientist-astronauts, though, the professional sacrifices NASA demanded of its astronauts were profound and damaging. In Houston, the scientists endured dismissals by their pilot-astronaut colleagues without even the barest of assurances that their sacrifices—career, family—would be rewarded. That so many of the Excess Eleven stayed on in NASA despite the repeated devaluing of their skills (becoming, in time, some of NASA's oldest and best-trained astronauts), was a puzzle to some. NASA, for its part, actively seduced the scientist-astronauts to stay even as it mistreated them. The agency arranged field trips to launch tests, for example, cultivating awe in the scientists even as it denied them flying opportunities. The scientist-astronauts—even the skeptical O'Leary—were smitten. He called it "launchitis, that euphoric disease which I temporarily caught when I saw that Saturn go up and which was sufficiently intense to keep me going in the program for the remainder of my six months in Houston."[161]

To O'Leary, though, entertainment was not a substitute for professional satisfaction, and the zeal of many of his colleagues to remain in NASA was inscrutable. "People in-house never wanted to admit their efforts were fizzling out. People out-of-house thought it was utterly stupid to be the fifteenth in line for a space flight in a program which had only two missions funded." Nonetheless, many of the scientist-astronauts remained devoted to their calling:

> The Excess Eleven remained optimists and I always questioned their optimism. . . . There was little agreement with my heretical pessimism—mostly surprise. I would probe the others about the weighty matters of risk, budget delays, the shoddiness of the Apollo Applications effort, the test-pilot dominance of the astronaut office, no science, flight school, living in Houston, and the likelihood of being away from home and family for more than half the time. Much to my regret, even the magnificent Excess Eleven seemed to ignore the most important negative aspects of the program.[162]

Perhaps, O'Leary concludes, "the drawbacks were too painful for their already made-up minds to consider."[163] Indeed, the longer scientist-astronauts stayed with NASA, the more difficult would be their transition back to academic life. When Apollo ended, physicist Joe Allen, like Curtis Michel and Philip Chapman, found himself so far removed from the scientific community that he was unable to resume physics research at his former level of ability, and as "an astronaut without a portfolio," served in NASA study groups devoted to post-Apollo human spaceflight before becoming—with some misgivings—NASA's assistant administrator for Legislative Affairs. For Allen, this seeming promotion threatened to end his abortive flying career.[164] For many, though, the promise of eventual spaceflight was sufficient to keep them in the corps.

For all of their determination, though, NASA's scientist-astronauts would not, as the scientific community hoped, restore basic science to prominence in American space exploration. Nor, though, were the men absent-minded professors likely to kill their colleagues with cockpit errors. The rigid structures NASA's pilot-astronauts had created to recruit, select, and train astronaut-candidates prevented such men from advancing, leaving the astronaut corps with a group of healthy, aggressive, and intelligent men no less skilled than any other junior military jet pilots. The barriers to the scientist-astronauts' participation in the space program were professional rather than technical. NASA had encouraged the men to apply by suggesting that they would, as the pilot-astronauts had, establish a new technical professional group, and many of the scientists who joined NASA retained some hope of doing so, despite mounting evidence to the contrary.

While NASA's pilot-astronauts had established a professional identity around their common expertise in flight testing, the scientists could not identify any common body of expertise upon which they could define themselves and which NASA valued. NASA's preference for physicians and lunar geologists suggested a specialized occupational niche in flight crews, but automation of experimental apparatus and a tendency to cross-train crew members in essential skills rendered the specialized knowledge of the doctors and scientists less critical for mission success. Rather than harness the men's unique talents, NASA's pilot-astronauts did what they had done with earlier generations of pilots: attempt to credential them in common flight test skills at the expense of their own individuality.

Classified as unqualified pilots instead of qualified scientists, most of the men languished. Astronauts selected in 1967 waited, on average, more than five times longer for their first flight that those of the 1962 group, and twice as long as the members of the 1966 group: an average wait of over 16 years. By comparison, NASA's Original Seven astronauts waited only five years, and the highly successful 1962 group, only three. Despite the often shabby treatment given them, scientist-astronauts, on average, enjoyed slightly longer careers within NASA than did the pilots, a surprising statistic considering that four of the 1967 scientist-astronauts resigned before they flew. When confronted with long wait times for their requisite two flights, scientists who did not leave immediately waited patiently and were rewarded, but the space program in which they flew little resembled the one they joined.

Chapter 4

The Man in the Gray Flannel Spacesuit

By 1970, the small cadre of veteran astronauts of the National Aeronautics and Space Administration (NASA) found themselves a dwindling force in a mammoth bureaucracy that had long since grown well beyond their control. That year, the astronaut-friendly Robert Gilruth departed as steward of NASA's human spaceflight program, leaving questions about the future of the effort, and about NASA's astronaut leadership. Meanwhile, the turbulent labor dynamic that had emerged with the arrival of scientist-astronauts intensified, as growing flight rosters, increasing layers of civilian management, declining budgets, and public scandals brought astronauts new occupational challenges. The early 1970s saw some of NASA's most dramatic successes in space—including the final Apollo flights to the Moon and the launch of the Skylab space station—but, for NASA's astronauts, the decade was one of diminished celebrity and autonomy, uncertainty about the future, and adjustments to a new kind of space workplace.

NASA's earliest astronauts had been the first to notice subtle changes in their workplace, beginning with the Apollo program. In the early 1960s, NASA's collegial, almost intimate management techniques had produced both success and tragedy; a sprawling new management structure adopted after the 1967 Apollo 1 fire left individual engineers—including flight crews—with less ability to influence hardware design and mission parameters.[1] Spaceflight was a socio-technical system of interconnected devices and professionals; after 1967, complex networks of overlapping review boards—"matrix management"—replaced small engineering teams, coordinating engineering and procedural changes to limit errors and incompatibilities.[2] For astronauts, the changes improved safety at the cost of their own authority.

Mercury astronaut Gordon Cooper felt sidelined in the new bureaucracy, later lamenting that an organization that "started with one hundred and fifty people involved in it" had expanded to an enterprise of "sixty thousand," in which minor changes that once took minutes to execute now took weeks. Cooper saw his fortunes fall in the new NASA, and retired in 1970. Cooper, up to his death in 2004, asserted that his performance in space actually established him as one of NASA's better astronauts, and accused a "government bureaucracy" of hamstringing him with arcane rules, paperwork, and tyrannical personalities. Assigned to one backup crew after another, Cooper suspected that veteran astronauts (especially Alan Shepard) were interested chiefly in securing plum flying opportunities for themselves. What remained of the Astronaut Corp's fragile meritocracy, Cooper insisted, disappeared when it came to assigning crews for the last Apollo flights. "I had been completely trusting," Cooper later wrote, "thinking they'd choose people on their strengths."[3]

To Wally Schirra, too, the era of seeming independence that marked the early, desperate years of the space race had ended by 1968, and his falling out with Deke Slayton, Chris Kraft, and Gene Kranz over Schirra's headstrong command of Apollo 7 that year was a cautionary lesson that other astronauts observed with trepidation.[4] During the 11-day Earth orbit mission, Schirra had protested an overloaded schedule and given his crew a day off against the wishes of ground controllers.[5] "I had always believed that I worked with NASA, not for NASA," Schirra later wrote. "By 1968 there was a bureaucracy developing—the fun days were over. I could see that I was out of line already." Schirra retired upon his return from Apollo 7.

These astronauts were not alone in leaving: having peaked in mid-1967 with 57 astronauts, the Astronaut Office saw its ranks drop to 49 in 1970 (despite the arrival of seven new pilots in 1969), and 28 in 1976. The image of the veteran spaceman—elite, reliable, imperturbable—would be a casualty of the 1970s, as space flights revealed astronauts to be fatigued and burdened by stress, fallible, and uncertain of the role upon their return to Earth. In the place of these former icons in the 1970s emerged a new cohort of astronauts: men of modest demeanor and even more modest hopes, for whom spaceflight was just another job, though one fraught with unique dangers and discomforts.

The MOL Transfers

Astronauts who arrived in NASA in the early 1960s had expected long careers capped by celebrity and interplanetary voyages. A small group of military astronauts transferred to NASA in 1969, though, found a different space program: shrinking, nervous, and consumed by organizational difficulties. In the late 1960s, military test pilots had sought opportunities to fly in space in at least three programs besides Project Mercury: X-15 and X-20 space planes, and the Manned Orbiting Laboratory (MOL). In the early 1960s, NASA's X-15 rocket plane had briefly crossed the threshold of space at suborbital speeds. The air force's X-20, a more capable replacement, was cancelled before flight, despite a strong early start. The same fate befell the air force's MOL in 1969; the program would have placed two-man crews of air force and navy astronauts aboard a reconnaissance space station beginning in 1972.[6] While the latter program in particular never flew actual crews into space, it did result in the recruitment and training of several military astronauts; pilots whose fortunes rose and fell with the air force's human spaceflight ambitions.

Unlike NASA's astronauts, the pilots chosen for these highly classified projects programs did not expect to become celebrities. Traveling the country under false names, in crew cuts and trench coats, MOL astronaut and air force officer Donald Peterson recounted, some in the public suspected they were intelligence agents.[7] Instead of spies, though, they were, like NASA's own astronauts, working pilots who had graduated from military academies or Reserve Officer's Training Corps programs at southern or western engineering colleges. While some possessed childhood interests in aviation and science, most did not expect to become professional pilots or spacemen, but found the work to be thrilling (and more appealing than other forms of military service) once exposed to it.[8]

For several of the men, graduation from the air force's Aerospace Research Pilot School preceded their selections; others had taken advanced degrees in science or engineering while continuing to fly. When the air force called Hank Hartsfield to active duty after ROTC at Auburn University in Alabama, he enlisted a professor to help him delay his service, as he had just begun graduate work in nuclear physics. Only later did he discover that he "loved flying."[9] Once credentialed as aviators, the future MOL pilots continued their scientific and engineering education while building their flight experience—a professional pathway that led directly to astronaut training. Peterson had bounced from one cancelled, high-profile program to another, gathering a master's degree in nuclear engineering, struggling to maintain his flying proficiency, and cultivating influential friends at Wright-Patterson Air Force Base while assigned to an intelligence unit.[10] Eventually, the men found themselves at Edwards Air Force Base, training or working alongside future NASA astronauts as eager as them, for a chance to participate in the space program. Unlike some of their earliest NASA colleagues, though, their introduction to space work came as a result of competition between military and civilian space programs, with the men struggling, weakly, to game a cryptic hiring bureaucracy.

For many air force and navy test pilots interested in spaceflight, it was not clear which of the various projects of the mid-1960s offered them the best opportunities, but the choice was seldom entirely their own. In 1965, only the MOL program hired new pilots; in 1966, both NASA and the air force sought pilots, leading many pilots scrambling to figure out to whom they should address their applications. Air force officers could apply to either or both programs, but those who applied for both, astronauts Hank Hartsfield, Charlie Duke, and Stu Roosa suspected, were prescreened by air force personnel and assigned to one or the other, to prevent the air force from finding itself in the awkward position of being turned down by one of its own in favor of NASA. The air force's Gordon Fullerton, a Caltech graduate and bomber pilot, had the choice made for him. Applying for both the MOL and NASA selections in 1966, Fullerton was assigned to the MOL selection and won a slot in the second class of MOL pilots.[11] Air force pilots who applied only to NASA competed with naval personnel for the available NASA positions.[12]

Naval officers at Edwards Air Force Base were also quickly swept up in the spaceflight boom. The navy's Richard Truly had intended to pursue a graduate degree in engineering when his squadron commander suggested the navy test pilot school at Patuxent River. Instead, he participated in an exchange program at Edwards, alongside later Apollo astronaut Fred Haise. Truly, believing himself to be unqualified for spaceflight, did not pursue the MOL selection that year, and was surprised when his name was entered into the pool simply for being a naval student at the air force's Aerospace Research Pilot School. The air force named him to the first group of MOL astronauts in 1965. (Indeed, Truly later remarked, "I'm the only person who has ever flown in space that never applied."[13]) Exploring a spaceflight career in 1966, Robert Crippen, another naval aviator and graduate of the Aerospace Research Pilot School, investigated both NASA and MOL, choosing to apply only to the latter. For Crippen, the decision was sheer calculation; even by the mid-1960s, "NASA had more astronauts than they knew what to do with" and Apollo faced an uncertain future.[14] MOL, by contrast, enjoyed DoD backing and had selected its first group of astronauts the previous year, seemingly offering a promising future.

Under the budgetary climate of the late 1960s, it was MOL, not Apollo, that lost its funding first. Three years into their training in the MOL program, Crippen, Fullerton, and their colleagues were abruptly informed of the project's cancellation; most heard the news on the radio rather than from superiors.[15] "[T]he MOL people were just crushed," Truly remembered. "I didn't know what was going to happen to my career."[16] Indeed, the MOL pilots realized that their participation in the highly classified MOL project prevented them from flying in combat overseas, closing off a career path that offered the next best route to promotion.[17] Without clear career plans, Bobko, Hartsfield, and Peterson chose to return to school and complete their military service.[18] Fortunately for many of the MOL pilots, though, the USAF had selected them for space duty while they were still fairly junior—many were under 30—and young enough to have a spaceflight career even after the MOL program folded. Manned Spaceflight associate administrator George Mueller decided to poach the "semi-trained astronauts" for NASA, providing a welcome option for the men who might have faced administrative duties. "Didn't have a lot of choice," Peterson recalled, faced with limited career options after MOL's cancellation.[19]

Slayton had neither needed nor wanted more astronauts in August 1969; when Mueller pressured Slayton to accept them, Slayton responded that he "didn't need some of the people [he] already had" and that no flights would be available for the MOL transfers in the near future. Gilruth, Slayton's superior, shared Slayton's assessment, but Mueller eventually prevailed, largely by arguing that accepting the MOL transfers would help NASA to secure air force support for future projects.[20] Touring NASA's Manned Spacecraft Center in 1969, the 14 MOL pilots received a hesitant welcome;

"a lot of the NASA folks didn't want us," Hartsfield remembered, but Alan Shepard had suggested vaguely that the proposed second and third Skylab space stations might offer flight opportunities several years in the future.[21] Ultimately, Mueller excluded, from the 14 MOL pilots, 7 who were over 36, and accepted the remainder into NASA: Karol Bobko, Robert Crippen, Gordon Fullerton, Henry Hartsfield, Robert Overmyer, Donald Peterson, and Richard Truly. Peterson and Hartsfield had just begun PhD programs under the air force's auspices, but withdrew soon thereafter, to the chagrin of their superiors.[22]

By the time the MOL pilot-astronauts reported for duty at NASA in 1969, plans for Apollo 18 and 19, and two proposed follow-on missions had already been cancelled.[23] For NASA's scientist-astronauts of 1965 and 1967, the MOL transfers further clogged the pecking order with potential crew members more qualified for crew positions than the scientists, even though they were junior to them, and often younger. Though arriving late in the Apollo program, the MOL transfers still belonged to the "fraternity" of military stick-and-rudder men who could be trusted to handle experimental vehicles. "[W]e just sort of slipped in there," Fullerton recalled, "they had a lot of astronauts," but many "were contemporaries of mine at test pilot school," or were "in fact…more junior."[24] Fresh off Apollo 11 and occupied with follow-up missions upon Fullerton's arrival, the Astronaut Office dispensed with the customary instruction that had so infuriated O'Leary. "They did find us a desk and an office," but "there was absolutely no training or indoctrination."

While recruited after their scientist colleagues, the MOL transfers entered a space program that valued their skills more highly than those of the scientists and offered them greater opportunities for early flight and command. In a space program that emphasized the proving-out of novel technologies, they possessed the background that ensured them interesting support work in the short term, and future opportunities greater than those offered to scientist-astronauts. Though the MOL transfers felt somewhat unwelcome, NASA's leadership still perceived these test pilots as possessing transferable skills useful in support work, inviting the new astronauts to pitch in where they could. "I returned decisions on operational hardware to those with a pilot background," Walter Cunningham recalled of his stewardship of the Skylab program, "depending heavily on the seven MOL pilots."[25] Fullerton soon learned to fly the Apollo Command and Service Modules (CSM), communicated with Apollo 12 crew members on the Moon, and closed the hatch on Apollo 17's Command Module (CM) just prior to its launch. He could scarcely believe his good fortune at receiving even this meager assignment.[26]

Cracks in the Facade

As NASA's newest astronauts settled into their jobs, its veteran astronauts became enmeshed in a series of scandals wrought by allegations of capricious management and financial improprieties.[27] After recovering from the vestibular disorder that had grounded him since 1963, Alan Shepard, in 1970, pushed aggressively for the next available Apollo flight, challenging Deke's Slayton's rotation system and causing considerable argument within the Astronaut Office. (That Shepard, then 47, was among the oldest of the astronauts, also attracted concern from the press.[28]) Slayton was accommodating to the fellow Mercury astronaut, but more junior astronaut Jim McDivitt—who would have flown as Shepard's Lunar Module Pilot (LMP) and who was concerned about Shepard's lack of training—threatened to resign if Shepard was assigned to his flight, and fellow astronauts were similarly displeased with Shepard's power play.[29] After protests from Mueller and McDivitt, Slayton bumped Shepard to the next mission, Apollo 14.[30]

With the return of Alan Shepard to the flight rotation, crew selection—always a controversial subject—became charged with allegations of naked favoritism and self-dealing by veterans at the expense of rookies. Aggressive lobbying by senior astronauts undermined morale among junior astronauts, causing them to question their leadership and bringing other longstanding labor practices into question. Among these were the private financial arrangements many veteran astronauts had cultivated while supposedly policing such abuses by junior colleagues. In 1971, David Scott's Apollo 15 crew smuggled 400 postal covers aboard their lunar module and later gave them to a German associate who agreed to retain them until the crew left public life, at which time they would be sold, partly to establish a trust fund for the astronauts' children. Instead, he promptly sold them though another middleman. The smuggling of souvenirs in spacecraft was commonplace in the space program, but a cash sale of flown materials grossly violated NASA rules. While the astronauts eventually refused the offer of payment, NASA management was nonetheless greatly displeased, and issued a formal reprimand to the Apollo 15 crew, who were promptly nudged out of the astronaut corps. It was this aspect of the controversy that was most galling to the men; they believed that as NASA's public face, they would enjoy lifetime employment if they needed it. "Astronauts don't get fired," Worden recalled thinking.[31]

Even this controversy, though, divided the astronauts, with some veterans siding with the Apollo 15 crew members and many junior astronauts criticizing the Astronaut Office as a weird fraternity loathe to properly reprimand veterans like Scott, who, Worden alleged, had orchestrated the scheme. A 1972 NASA investigation of the astronauts found that a small number of astronauts had engaged in the trafficking of signatures and other forbidden commercial activities, though some (like the Apollo 15 crew) were likely duped by middlemen seeking to parlay astronaut signatures into quick cash. Other veteran astronauts, though, studiously avoided such scams, and responded angrily to the suggestion that they had compromised NASA's reputation in any way. Veteran Apollo astronaut John Young, in particular, angrily denounced NASA Inspector Ray Wood, denying any wrongdoing and criticizing the agency for requiring astronauts to submit to such quasi-legal questioning.[32]

Despite such protests, though, the stamp controversy brought unwanted publicity to the Astronaut Office and unprecedented scrutiny of its inner workings and leadership. Scientist-astronaut Donald Holmquest wondered if veterans like Shepard and Slayton had grown complacent, lazy, and even "corrupt" in their management of their fellow astronauts. "I feel strongly," Holmquest wrote in response to a 1972 NASA interrogatory, "that the management of the astronaut office has rested too long in incompetent hands, with repeated violations of common sense and misuse of government equipment overlooked." Yet, while the Apollo 15 controversy had shed light on certain embarrassing practices within the Astronaut Corps, even Holmquest was unwilling to see Scott "publicly sacrificed" for what he considered to be leadership failures endemic to the organization. "In my opinion, NASA has exercised extremely poor judgment in permitting such absolute power over events to be held by poorly trained and motivated persons such as Mr. Slayton, Stafford, and Shepard," Holmquest wrote. Such men had turned a blind eye toward not only astronaut self-dealing, but also to Shepard's smuggling of a golf club head and ball to the Moon on Apollo 14, certain astronauts' poor safety records, and, presumably, the continued ostracism of the scientists. So incensed was Holmquest that he threatened to "make these and several other violations of the public trust a matter of public record" if "Mr. Slayton, Stafford, or Shepard are chosen for the Russian docking flight [the 1975 Apollo-Soyuz Test Project (ASTP), discussed below], or perhaps if they escape public reprimand."[33]

Rather than mollifying critics, Slayton, incensed by the attention lavished on the astronauts' private work culture, lashed out angrily at reporters. In one damaging 1972 article in Tomball, Texas's *Tomball Tribune*, an intemperate and profane Slayton not only appeared to defend the astronauts' practice of carrying memorabilia into space, but suggested that "every crew" to land on the Moon had stripped their Lunar Module (LM) of dials, switches, and other mementos, and possibly smuggled Moon rocks home as well. Instead of condemning such behaviors, Slayton angrily dismissed the idea that he strip astronauts' "bare-ass" upon their return in a vain search for contraband. Rather, Slayton asserted that he secretly maintained a list of everything the astronauts claimed to have brought back to Earth, an inventory he considered "private" and would not reveal. The article made rounds among top NASA officials, including Fletcher, who forwarded it to Low with a note suggesting that Slayton "be instructed to stay away from reporters for an indefinite period," a request to which Low assented.[34] For the first time, NASA had questioned Slayton's ability to continue in his administrative role, and had begun to look more closely at the astronauts, whose labor disputes were finally becoming public.

The following year, Holmquest, in a letter to NASA administrator James Fletcher, again singled out Shepard, Slayton, and Stafford for manipulating the space program to serve "their own interests and personal gain."[35] Several of the astronauts had benefited materially from signing stamps or lugging souvenirs to the Moon, including Shepard, who was ostensibly responsible for policing such abuses.[36] Holmquest wrote that Slayton had been, for the past several weeks, attempting to force Holmquest out of NASA, and suggested that other scientist-astronauts would soon follow. Holmquest, at the end of a NASA-approved leave of absence, wished neither to leave the astronaut corps nor return to NASA only to find himself subjected to the "prejudice" that had been inflicted upon him earlier.[37] Holmquest's protests, though, yielded no observable changes in the Astronaut Office. Both Slayton and Stafford flew in space again, while Holmquest never followed through on his threat to expose the program, and resigned in 1973 having never received a prime crew assignment.

Amid such departures, unflattering press reports about the astronauts' activities continued through the early 1970s. During the 1960s, astronauts inclined to criticize NASA or their colleagues seldom aired their concerns. With Apollo nearly over, though, NASA could no longer use flight assignments as a stick to induce cooperation. In 1972, *Life* chose not to renew its publicity contract with the astronauts, ending their period of automatic celebrity, reducing their earnings, and giving them greater freedom to vent their dissatisfaction in print. Unflown scientist-astronauts astronauts Brian O'Leary and Curtis Michel were the first to expose NASA's office politics, but the flurry of newspaper and magazine articles that followed further highlighted embarrassing aspects of astronaut culture and cast veteran spacemen as hustlers abusing the public trust.[38] Publicly, junior astronauts lashed out angrily against Slayton's perceived cronyism, while veterans accused NASA management of undermining their professional autonomy even as they scrambled to fill the last of Apollo's flights. These issues came to a head during the flight of Skylab, America's first space station, which brought NASA's astronauts into public conflict with management over a host of labor issues, from personal privacy to work schedules.

Labor Unrest on Skylab

While NASA remained focused upon achieving lunar landing, controversies surrounding astronaut labor often attracted little attention in the press. That the astronauts would land on the Moon was sufficient to motivate the astronauts to tolerate a certain amount of internal dissension and conflict with NASA management. With Apollo lunar missions over in 1972, though, astronauts quickly adapted to a postlunar workplace in which spaceflight no longer garnered the same degree of public attention, and even assumed an appearance of routine. This appearance was deceiving: in Apollo flights to the Skylab space station, astronauts encountered new challenges as they attempted to adapt their work dynamics to spaceflights lasting months instead of days. Skylab missions, flown during 1973 and 1974, posed challenges not of the dangers of lunar landing, but the less glamorous mechanics of crew selection and the rigors of managing complex work schedules in space. Conflicts in 1971 and 1972 over the crewing of Skylab missions of 1973 and 1974 were particularly heated as astronauts scrambled for what they perceived to be the last flying seats many would experience in their careers. As the scientist-astronauts fretted over their diminished role in Skylab, pilot-astronauts at first struggled to avoid what they initially considered to be an unglamorous assignment, and then pushed aggressively for what appeared to be some of the few remaining slots in the rapidly declining Apollo program.

When NASA had established the Office of the Apollo Applications Program (AAP) in 1968, its management had shuffled between a number of astronauts without prime crew assignments—Alan Bean, Gordon Cooper, Owen Garriott, Walter Cunningham—most of whom feared that duty with this the low-priority project had sidelined them for lunar missions. Their concerns were reflected by the staff NASA had assigned to assist them—virtually all of the scientist-astronauts, and NASA's most junior pilot-astronauts, the MOL transfers, whom no one expected would fly for several years. Instead of complaining, though, the men accepted the duties offered them while hoping for lucky breaks that never came.

NASA intended Skylab missions to be led by veteran pilot-astronauts, but by 1972, an insufficient number remained to staff backup crews. Only when Skylab's plan solidified in 1970 did a senior astronaut, Pete Conrad, take charge of both Skylab's management and its first crew, with Alan Bean commanding the second Skylab mission and an all-rookie team crewing the third. The mixture of veterans, rookies, and scientists displeased virtually everyone associated with Astronaut Office: Slayton, who wanted more veterans; Homer Newell, who wanted more scientists, and rookie astronauts, who wanted more of themselves.[39] Veteran Walter Cunningham, having managed the AAP upon his return from 1968's Apollo 7, hoped to secure a command position on a future Apollo flight, though he lacked experience in space rendezvous and docking (NASA's preference for Skylab crew commanders).[40] Slayton's schedule left him backing up Conrad; shunted first from his management job and then a prime crew slot, Cunningham resigned from NASA in January 1972.

That year, two MOL transfers, Robert Crippen and Karol Bobko, and scientist-astronaut William Thornton conducted a ground simulation in a Skylab mockup—the Skylab Medical Experiments Altitude Test (SMEAT)—a 56-day marathon that left George Low wondering if NASA didn't owe the men at least some kind of commendation, even if they did not actually go anywhere.[41] Most, like Truly, enthusiastically approached ground work on Skylab in the hope that they might get an opportunity to fly to the station. "I foolishly thought that maybe I would actually get to fly on Skylab," Truly recounted, "it wasn't to be."[42] MOL transfer and air force pilot Hank Hartsfield found himself stowing repair equipment in Pete Conrad's Apollo CM minutes before the first piloted Skylab mission, accidentally kicking Joe Kerwin in the stomach and wishing he could stow away in the capsule for his first flight into space.[43] These support activities, though, were the closest to flying the men would encounter in the Skylab program.

Upon its launch, unpiloted in 1973, Skylab was beset with immediate problems, including damage to the station's micrometeorite shield and two large solar power arrays. After ground personnel identified the necessary repairs, two members of the first Apollo crew to visit the station, Pete Conrad and scientist-astronaut Joe Kerwin, clad in spacesuits, eventually restored Skylab to habitability. Successful in making the station operational, Conrad, Kerwin, and Skylab 2 crewmate Paul Weitz remained onboard for 28 days. As a space laboratory, Skylab eventually proved adequate to the tasks assigned to it. The abundant internal volume of the converted fuel tank provided the astronauts with the internal volume equivalent to that of a small, two-story house, with a galley, sanitation area, exercise room, airlock, and a variety of scientific apparatus, including a solar telescope. A large deck in the center of the station provided enough volume to enable the astronauts to somersault freely without touching the station walls, a feature crews exploited to test jet packs and engage in exercise and recreational activities.

The astronauts, though, had not been launched in space for entertainment purposes; a teletype printer in the station radioed, to the astronauts, daily schedules of tasks associated with the station's various scientific objectives, including solar astronomy and physiological research. Each evening, ground crews, consulting with each experiment's principal investigator, transmitted a schedule for the next day's activities, often leaving the Skylab crews exhausted. The amount of work a space crew could perform in space remained an open question throughout the program, as NASA adjusted to the working pace of different crews and the logistical demands of a new space station suffering from chronic power shortage and unexpected maintenance problems. Conrad's Skylab 2 crew had labored hard to meet ground expectations for scientific output but were ultimately unable to do so, despite giving up some of their recreation time. The Skylab 3 crew enthusiastically exceeded their work quota, sacrificing communal eating and exercise time to complete more work.[44]

Commander Alan Bean's Skylab 3 crew enjoyed a 59-day stay, accommodating to their new environment and performing valuable scientific research. Skylab, though, was still a potentially hazardous environment. Failing gyroscopes had to be serviced on the exterior of the station, a procedure NASA had not expected, but which the astronauts completed successfully. And when the crew's Apollo vehicle developed problems with its maneuvering system while docked to Skylab, NASA became concerned that the spacecraft would degrade further and become unmaneuverable, stranding the astronauts in the station until their oxygen ran out. In a feat worthy of a science fiction novel, NASA hurriedly fitted out a rescue version of the Apollo CM, complete with five crew couches for the three Skylab 3 crew members and astronauts Vance Brand and Don Lind, who would fly the Apollo spacecraft to the station to rescue the Skylab 3 crew. NASA though, with the help of Brand and Lind, was ultimately able to isolate the problem with the Skylab 3 CSM, and the rescue mission was never flown.[45]

The third crewed mission, Skylab 4, not only further extended the record for American long-duration spaceflight but demonstrated that, as with Apollo 7, isolated astronauts under pressure to perform might rebel against ground controllers, if overworked. For the first time since the 1966 Gemini VII mission, NASA flew an all-rookie crew—not a problem for the well-trained astronauts but a concern for NASA physicians unsure how the men would react in space. Some, crew member and scientist-astronaut Ed Gibson recalled, feared that a publicized bout of space-sickness might jeopardize the future of NASA long-duration human spaceflight program, and insisted that the astronauts medicate themselves for nausea prior to flight. The astronauts, concerned that the medications would make them too dizzy to safely abort their launch in case of trouble, preferred to wait until the onset of symptoms before ingesting the prescribed drugs.[46] To Gibson, the opinion of NASA's medical community was divided and unreliable; certain physicians concerned with their reputations and that of the program were exercising undue caution, while the "rational physicians we worked with every day" thought otherwise. Confronted with a "decision making process" they could "no longer trust," the astronauts secretly failed to premedicate, leading CMP Bill Pogue to become ill early in the flight.

Recognizing the trouble they were now in, the astronauts resolved to bag the effluence for scientific analysis, but not to inform the ground until their return. Unfortunately for the astronauts, audio equipment monitoring cabin communications recorded the astronauts' deliberations; the tapes, when dumped to ground stations, informed the Mission Control Center of the deception. Shepard chastised the crew; "a hell of a way to start a flight," Gibson recalled.[47] Recalling the event many years later, Gibson regretted that he did not lie to the ground merely about taking the pills. It was the first of several difficulties encountered by the crew meeting the expectations of ground teams. Throughout the 84-day visit, Skylab's crew chafed against what they perceived as burdensome responsibilities and unreasonable intrusions on their personal lives.

Entering the dark, vacant space station, the Skylab 4 crew found three dummies waiting for them, assembled from discarded clothing by the station's previous occupants. While claiming to enjoy the joke, the astronauts were so pressured to keep on schedule that they did not have time to dispose of the dummies immediately, and found their ongoing presence on the station disconcerting. Morning routines included daily weighing, close monitoring of mineral consumption, and the collection and bagging of their urine and feces for later analysis. (Later testing revealed that the astronauts had lost bone density in orbit). Having food aboard the station enough for only 56 days of operations but hoping to extend their visit even longer, the astronauts resolved to consume only processed food bars for breakfast every third day, "one of the most supreme sacrifices anyone has ever made for the space station or the space program," Gibson recalled. "[Y]our breakfast consisted of four or five crunches and that's breakfast. . . .I still have a tough time looking at a food bar in the face now."[48]

Communications, conducted on unscrambled, open channels available to the press, continued to vex the men for the duration of their flight. Private conversations between either the astronauts or astronauts and the ground required use of a private channel and were discouraged, leading the astronauts to avoid discussing controversial issues for fear of exposing or embarrassing themselves.[49] Chief of these issues were medical concerns and the mission's ambitious experimental schedule. Determined to exploit Skylab's longest flight, ground controllers increased the experimental workload of the third Skylab crew, conveying up to 60 feet of instructions per day, Gibson recalled. These experiments usually required little more than that the men set switches and dials, Gibson recounted; "it never gave you any time to really use your intelligence in how you took data. It was just push the buttons as fast as you can and move on to the next."[50]

Productive in their work but falling behind an overambitious seven-day-a-week-schedule, Commander Jerry Carr requested Sundays off; as in 1963 and 1968, the request eventually created a popular impression that astronauts were attempting to wrest control of their workplace from NASA management. The later suggestion that the astronauts had actually "mutinied" in space, though, offended them. Consumed by an experimental backlog, Gibson wanted not a day of rest, but a day to catch up on his schedule, with no new instructions to occupy their time. "I always worked every day," Gibson noted, "I was going to work Sundays anyway." Compounding the apparent insubordination, though, the astronauts, Gibson later claimed, accidentally deactivated their radio, and, when they reactivated it, realized that they had missed communications from ground controllers. It is unclear exactly what occurred, but the image of rebellious astronauts proved irresistible to the press, which wrote of the event as "strike in space," a characterization that grated on Gibson. Reality, he suggested, was far simpler. "[W]e screwed up," Gibson explained.[51] Eventually, though, Flight Director Gene Kranz relented, giving the crew several days off, which they spent working on the station's solar telescope.[52]

Skylab's three successful piloted flights and surplus hardware promised a string of successful, relatively inexpensive space missions that would launch NASA's remaining astronaut roster into space relatively quickly, but the program suffered from budgetary limitations intended to preserve NASA funds for follow-on vehicles. By 1974, Apollo had proven that NASA's existing hardware was flight worthy, safe, reliable, and relatively easy to modify to meet new operational objectives. It pointed to a future in which expendable heavy lift rockets would blast large payloads into space, to be followed by small piloted ferry vehicles made cheaper through efficient design and mass production. Skylab demonstrated that NASA could modify Space Race hardware to accomplish good science, while the flights of three more of NASA's scientist-astronauts validated the concept of the Science Pilot. Instead, human occupancy of Skylab ended in 1974 with the return of Carr, Pogue, and Gibson; future Apollo visits to Skylab were not forthcoming. Originally intended as a plan of up to five three-man flight, followed, possibly by the orbiting of a second space station between 1975 and 1977 and a third by the early 1980s, Skylab eventually repeated the disappointments of the lunar program, including a premature termination with a large roster of rookie astronauts who would have likely flown these follow-on missions.[53]

The abandonment of Skylab and the end of Apollo flights provided an inducement for several more veteran astronauts to leave NASA. On his return from the Skylab 2 flight, Pete Conrad resigned following a tense exchange with NASA managers over the leadership of the Astronaut Office during the previous two years. Announcing his impending retirement to NASA, Conrad had placed an angry telephone call to NASA administrator James Fletcher on October 17, 1973, in which Conrad criticized in particular NASA's seeming distrust of its astronauts in the wake of the Apollo 15 stamp matter. NASA had demanded written assurances from astronauts that they would not transfer space program artifacts outside their families without first informing NASA, a new requirement Conrad found demeaning. This slight, though, was only the latest in what Conrad had considered to be a series of petty indignities he and his fellow astronauts had suffered at the hands of NASA management since 1972, including NASA's continuing efforts to bar astronauts from wearing their service uniforms in public.[54] As a pretense for resignation, though, this longstanding rule was a hollow justification. "The final straw," Fletcher later recounted to George Low, "was what Pete called the breaking up of the Astronaut Office, which he again attributed to me personally."[55]

For scientists-astronauts who had waited for seats on Apollo, NASA's eventual abandonment of Skylab was an equally bitter disappointment. To Philip Chapman, who had left NASA in 1972, Skylab was doomed by its failure to glorify NASA's pilot constituency: orbiting Earth unpiloted, it offered NASA a platform for scientific work in which the NASA's astronauts served little function. Even Skylab veteran Owen Garriott, who seemed loathe to publicly criticize NASA human spaceflight efforts, wondered why NASA had not crewed Skylab flights with two scientists instead of one, or why it abandoned a second completed Skylab Orbital Workshop to museum use. This "calamity" eliminated an obvious opportunity to make good on NASA's promise to fly at least some of the 1967 scientist-astronauts in the Apollo vehicles they knew best.[56] The original Skylab, empty, remained in Earth orbit until 1979, in part because NASA could not devise a safe plan to deorbit the space station for a controlled reentry.[57] A second, unflown Skylab eventually found a home in the Smithsonian Institution's National Air and Space Museum, where, reputedly, several NASA astronauts refused to visit it, still stinging from the disappointment of its grounding.[58]

Last Flight of Apollo

The conclusion of piloted flights to Skylab ended the hope of most of NASA's astronauts for immediate flying opportunities. For the final flight of Apollo, a joint mission flown with the Soviet Union in 1975, veterans again jealously guarded the few remaining flight seats and shunted rookies to a support role. In a test of international rendezvous techniques, an American Apollo crew would maneuver toward and dock with an orbiting Soviet Soyuz spacecraft and engage in joint scientific research with the Soyuz's two-man crew. While the Apollo-Soyuz Test Project provided opportunities for some of NASA's oldest astronauts to fly, the men's checkered performance on the flight accelerated doubts about the reliability of the veterans who remained in NASA's astronaut corps.

In crewing the ASTP mission, Deke Slayton, scouring NASA's roster, found few of America's veteran astronauts willing to take assignment and too many rookies to possibly accommodate them all. Ultimately chosen to lead the mission was veteran astronaut Tom Stafford, who had so much clout within NASA that he had turned down command of the last Apollo flight to the Moon, having already orbited on Apollo 10. Command Module Pilot (CMP) Vance Brand had waited for a seat for nine years as the Apollo program rose and ebbed. Slayton, in his first spaceflight, claimed the new "Docking Module Pilot" seat; his last duty as Chief Astronaut was to assign himself to the last American space crew until the 1980s (Figure 4.1).[59]

Slayton, like Shepard, found a cool reception after returning to flight status in 1972; though he would fly as the third member of the crew, it would not be in the position his seniority would normally dictate. Rather, NASA insisted that experienced 1962 group astronaut Stafford command the mission, rookie Brand fly as CMP, and Slayton occupy the lowest-ranking seat, a position that required little actual flying. Slayton was disappointed—he was,

Figure 4.1 The crewmembers for the 1975 Apollo Soyuz test project included American astronauts Donald "Deke" Slayton, Tom Stafford, Vance Brand, and Soviet cosmonauts Aleksey Leonov and Valeriy Kubasov (NASA photo)

at the time, the senior-ranking NASA astronaut and preferred a more prominent role on the flight—but he accepted NASA's decision.[60] Again, junior astronauts were furious at the apparent selfishness of senior colleagues. Walter Cunningham resented Slayton's assignment of himself to the flight when Cunningham had faithfully stewarded the program years earlier.[61] MOL transfers Karol Bobko and Bob Overmyer were eminently qualified, having spent years preparing to spy on the Soviet Union as military astronauts in the air force's MOL program. Instead, the men found themselves among the members of the support crew for the flight, learning Russian and touring Moscow. Walking through Red Square, Overmyer was circumspect, telling Bobko that, he "never doubted" that he'd visit the landmark, but that he'd always expected "it would be at 200 feet and a full afterburner."[62]

Launch of Apollo-Soyuz on July 15, 1975 proceeded without significant problems, and was soon followed by the interception of the Soviet craft and its crew, Aleksey Leonov and Valeriy Kubasov. When the Apollo spacecraft left Earth orbit and reentered the Earth's atmosphere, though, potentially life-threatening problems began to emerge. During the craft's descent, failure of the crew to flip a switch shutting down the spacecraft's rocket thrusters resulted in toxic fumes entering the Command Module, a compound oversight for which Stafford took the blame, but for which all crew members were partly responsible. Brand held the critical responsibility of "flying the reentry," but either Brand had missed Stafford's order to flip the switch, or Stafford had never given it, and neither caught the error.[63]

As Stafford later described in his memoir, the capsule struck the ocean and immediately flipped upside-down into the "Stable II" configuration, which submerged the escape hatches and left the choking astronauts strapped to ceiling. Stafford unhooked himself, falling down into the nose of the capsule and injuring his shoulder and elbow. He scrambled to inflate the capsule's flotation bags, and handed oxygen masks to his crewmates. Brand, though, was already comatose, his hands clenched. Stafford struggled to put a mask over Brand's face; Brand, suddenly awakened and flailing violently, punched Stafford in the face, knocking Stafford into the instrument panel. Stafford continued to struggle with Brand, who again lost consciousness as the mask slipped off. Stafford grabbed hold of Brand and again revived him; by then, though, the masks were out of oxygen, and the cabin was still filled with fumes that burned the astronauts' faces and throats.

Radio problems had prevented the astronauts from communicating with recovery crews throughout the splashdown phase; as frogmen approached, the astronauts remained in grave danger. Stafford reached to open the hatch to vent the fumes, but Deke stopped him. "Goddamn, don't open the hatch!" Stafford recalled Slayton shouting. "We might sink like Gus!"—Gus Grissom, Slayton's best friend in the astronaut corps, who, 15 years earlier, had nearly drowned after splashdown when water flooded through an open hatch on his Mercury capsule. Stafford was struck by the comment; at the time of the ASTP flight, Grissom had been dead for seven years and Slayton appeared to be panicking.[64] When recovery crews finally arrived, Slayton mistakenly signaled to the frogmen that the crew was in good health. Stafford, when finally able to escape the capsule, insisted that his crew stay put instead of being dragged from the water like Grissom had been, preferring that the spacecraft be hoisted aboard ship and the crew disembark with dignity.

Shepard's, and, later, Slayton's, seniority and test pilot qualifications enabled them to overcome concerns over their relative lack of recent training, but doubts over their ability to command missions after years spent in desk jobs clouded their return to flight. All three men aboard the ASTP flight were experienced test pilots, but some questioned whether Stafford and Slayton had trained sufficiently. "Forty-four-year-old rookie" Vance Brand bore a disproportionate share of the responsibility for the flight, especially for flying the vehicle during its reentry to Earth. Walter Cunningham, then retired from NASA, placed the blame for the ASTP accident on the inadequate training regimen and "casual approach" of "old pro" Stafford and "forty-four-year old rookie[s]" Slayton and Brand. Cunningham's original 1977 account of the incident concluded with a judgment about the perils of flying older astronauts:

> Flying a spacecraft is really not all that difficult or demanding. There is certainly no obvious reason why a healthy, well-trained, forty-five- or fifty-year-old pilot cannot safely handle it. The training grind is a different story! It is a hectic, physically and mentally demanding pace for anyone, and especially trying on forty- and fifty-year-olds.[65]

NASA, Cunningham writes, "considered adopting a higher degree of automation" after the mishap.[66] Indeed, the troubling performance of the Apollo-Soyuz crew added to the concerns that already surrounded the mission.

Fears that space vehicles had finally grown too complex for even NASA's veteran pilot-astronauts continued well into 1978, when Chester Lee, director of operations for the new space shuttle program, drafted a letter to various NASA managers concerning revisions to astronaut training procedures. In the light of "our ASTP landing experience," Lee wrote in August, astronauts would "not be permitted to either change or deviate from well established procedures," a requirement NASA had never before put in writing. The letter's recipients included Slayton himself, whose performance was under scrutiny, and who was now supervising space shuttle development in the hope of obtaining an early pilot seat. "[A]fter listening to and analyzing the voice tapes" of the mission, NASA investigators had concluded that "any deviations would have to be approved by some authority higher that the crew commander," an indictment of Stafford's leadership.[67] While crew commanders had always sought to follow the flight plan and followed instructions from Mission Control, Lee's declaration was a symbolic blow to the delicate appearance of authority the astronauts had previously established.

ASTP marked the passing of both a generation of veteran spacefarers and the ideal of flawless performance that the earliest crews had cultivated. The mythology of Apollo—that any astronaut selected could fly any mission at any time—was shown to be faulty. NASA's success was built not only upon the men it had selected but on an intensive training regime and rapid operating tempo that maintained flight personnel in peak condition. Absent frequent flights and a cadre of such skilled astronauts, certain NASA managers began to wonder if space vehicles might be better run by computers, tended by men who followed orders. The MOL transfers, still awaiting their first flights, would be those men.

Between Apollo and Shuttle

NASA's seven MOL transfers (and most of NASA's remaining scientist-astronauts) would find their first spaceflights aboard the space shuttle, an Apollo replacement NASA had studied during the late 1960s as an element in a broad array of space exploration technologies proposed for the 1970s. The shuttle, a winged orbital space plane the size of a small jet airliner, would be larger than previous space vehicles, with more room for both pilots (who would fly the vehicle) and a large complement of scientist-astronauts, the shuttle would ferry into space for weeks at a time, dozens of times a year.[68] As military pilots, the MOL transfers would be integral to the development and testing of the space shuttle, the only piloting opportunities left in the space program by 1975.

Though intended to democratize space travel, the shuttle was a pilot's dream. While the vehicle would fly automatically for most its flight, landing required human piloting, helping to ensure that no scientist-astronaut would ever command or pilot a shuttle mission. For astronauts assigned to the shuttle, space work was much closer to traditional test piloting than the experience of Mercury, Gemini, or Apollo crews. The technically adept MOL transfers slipped almost effortlessly into the work, flying test vehicles and chase planes, diagnosing systems problems, and creating new procedures, including, in the case of Peterson, a technique to fly the shuttle into orbit manually if the guidance system failed.[69] Unlike America's previous space vehicles, the space shuttle would launch for the first time on a piloted orbital flight, challenging the vehicle's subsystems to work correctly on their first sustained test.

After its launch via rocket boosters and its own liquid-fuel engine, the shuttle Orbiter would glide to a piloting during landing, a challenging maneuver that, in the unpowered craft, would offer little margin for error and would require careful simulation before flight. Upon its roll-out in 1977, the shuttle Orbiter *Enterprise* soon underwent captured flight and drop-testing from the back of a converted Boeing 747 jetliner, with piloted landings flown by two crews, each consisting of one more senior pilot-astronaut and one MOL transfer (Fred Haise[70] and Gordon Fullerton; Joe Engle and Richard Truly). Working on the shuttle's cockpit instrumentation and later flying Approach and Landing Tests (ALT), Fullerton found himself not merely training to fly the shuttle, but establishing procedures and techniques that would need to be refined and practiced for all subsequent crew members (Figure 4.2).[71] Indeed, the shuttle's instrumentation was filled with counterintuitive gauges and flaws that went unremedied due to cost concerns. The MOL transfers fretted quietly over how to fly the shuttle given its quirks; successive astronauts, Donald Peterson later noted, may have never even learned about them.[72]

Less glamorous than spaceflight but even more demanding, Fullerton recalled, early shuttle testing required countless practice flights in other aircraft modified to behave like the shuttle in flight, tedious work that would be absolutely essential in ensuring the safety of later crews. Flying chase in a T-38, Karol Bobko struggled to make his nimble jet fly as slowly as the lumbering space glider, "shaking the stick to try to get a little bit more drag into the T-38 to have it not overrun the Shuttle."[73] The shuttle Orbiter was projected to handle poorly and pilots frequently overcorrected on the flight controls, leading to oscillations and, eventually, the occasional hard landing. In an interview for 1975 *Time* article, Bobko provided a rare damning indictment of the vehicle, declaring that the shuttle "must have been

Figure 4.2 Commander Fred Haise (left) and pilot C. Gordon Fullerton pose with space shuttle Orbiter Enterprise at Rockwell International Space Division's Orbiter Assembly Facility at Palmdale, California, in preparation for Enterprise's 1977 approach and landing tests (NASA photo)

designed by a brick mason," and asking, "If the wings fell off would the pilot even notice it?"[74] Shuttle astronauts surmounted these problems through rigorous preflight training, including simulator work.[75]

Much of Fullerton's time was, like Crippen's, devoted to computer systems analysis and hardware and software debugging; as in much of commercial America in the late 1960s and early 1970s, computers were rapidly assuming a critical place in space vehicle control and navigation. The curriculum at the air force's Aerospace Research Pilot School notably included a course on digital computing; computers not only ran the shuttle's navigation system, but its flight controls as well, though a fly-by-wire system that translated movements of the control stick and pedals to control surfaces on the wings and rudder. In the *Enterprise*'s right seat for the first of its approach and landing tests (where he would monitor the ships systems) Fullerton recalls, a computer display failed almost immediately. So consumed was he with a recovery checklist that he barely noticed that the shuttle could actually fly.[76] Computer issues consumed the working hours of several astronauts, including Peterson, who wrestled with issues of redundancy in the shuttle's bank of five computers, any three of which had to work to ensure the shuttle's ability to navigate.[77]

Increasingly, astronaut work required close interaction with computer systems; as in private industry, software proved a major bottleneck.[78] Elsewhere, other revolutions in computing seemed to find parallels in the design of the shuttle. Interaction between the flight crew and the shuttle's computers was largely numerical, with teams of ground controllers always on standby to help interpret the data. Peterson lobbied (unsuccessfully) for a more intuitive, more personal graphical interface that would enable shuttle pilots to fly a reentry when out of communications range of ground stations.[79] Unwilling to make such an expensive change, though, NASA continued to rely upon the more rudimentary systems, eventually supplemented by off-the-shelf laptop computers. Often, computer problems proved so intractable that they went unrepaired, as doing so would require the laborious requalification of flight hardware already approved for use. Early shuttle flights flew with a book entitled *Program Notes*, containing known system bugs, including one that caused a display to read "off" when it was supposed to read "on."[80] Like many other technical workers of the late 1970s, the MOL transfers eventually thrived in NASA by learning to operate its computers and not complaining too much.

Eventually, these men emerged as NASA's senior pilots, stewarding crews into space well into middle-age, but for the time being, their job combined all of the difficulties of the astronaut life with none of the perks. In pay, especially, they lagged behind former Apollo crew members. While earlier-established NASA compensation policies had ensured that astronauts would received salaries commensurate with their experience, they privileged astronauts who had flown in space over rookies, regardless of the rookies' tenure. Meanwhile, as more astronauts joined the program, publication royalties, which were distributed evenly among the astronauts, dwindled to an inconsequential sum. Field allowed the contracts to expire in 1967 and *Life* in 1970, just as the MOL transfers were settling in to NASA. By 1977, the MOL transfers, in training as astronauts since the mid-1960s, had, described Flight Director Chris Kraft, suffered "severe salary inequities," that required an adjustment "to set astronaut salaries based on a total evaluation of the candidate's experience, background, and training."[81]

Meager increases in their government salaries, though, would not compensate for the fact that shuttle astronauts, while able to participate in challenging and rewarding flight test work, would likely never become celebrities on the order of the Original Seven. The early test pilots at Edwards had never expected fame; the professional currency in which they traded was the respect of their superiors and their positions in the flying hierarchy, a highly defined pecking order that determined who would fly what, and when. The loss of celebrity status that astronauts experienced in the 1970s returned them to their former status. The small community of spacemen increasingly assumed the character of a corporation organization run by distant managers, in which pilots served as anonymous technicians, secure, but unheralded. Joining NASA during its decline, the MOL transfers developed a different set of expectations accompanying their work, and settled into a professional identity rooted less in celebrity than in the slow, careful, and often dangerous slog of test pilot work.

While program managers of the 1950s and 1960s obsessed over which humans, if any, possessed the "right stuff" to fly in space, by the end of the 1970s, astronauts like Deke Slayton worried whether the shuttle was an adequate successor to Apollo. The lack of abort modes throughout the shuttle's predicted flight plan was a constant concern for astronauts training to fly it; contingency plans struck the pilots as half-formed and unrehearsed. Proposed abort techniques during the shuttle's ascent, Donald Peterson recalled, were "just unbelievable," requiring delicate maneuvers that had never been performed: "You're flying outbound and you're inverted. If you have to come back to the launch site, the first thing you do is pitch the vehicle over, and then you're flying backward, and you're flying backward for a long period with the engines burning to slow you down and start you back in. ...When you get to that target line, essentially what you do is you pitch the vehicle down pretty violently and you shut off all the engines and you dump the tank, and you do all kinds of very dynamic maneuvers that you don't do in a normal flight. None of that has ever been tested."[82] Unlike Apollo, the shuttle would be "man-rated," with its very first test flight, an unheard-of acceleration in normal testing procedures.[83]

Protesting the program would do the astronauts little good, and constant memos circulated by new Chief Astronaut John Young about the latest problems became a well-known joke within the Astronaut Office.[84] Indeed, the influence of the pilot-astronauts, which Cunningham credited for the space program's impressive safety record, had already begun to dissipate in vehicle design:

> A fair amount of the credit for the operational success and the safety records of they Mercury, Gemini, and Apollo Programs was due to the astronauts and

> their contributions to systems design and operational planning—the historic test pilot's role. All of the early astronauts were tough-minded test pilots and fighter pilots who refused to fly a vehicle that didn't reflect their operational viewpoint. NASA management, acknowledging the experimental nature of the programs, was quite receptive to astronauts' views for the Mercury and Gemini Programs, and, to a lesser degree, for the Apollo Program.[85]

Mercury and Gemini had been built with substantial astronaut input, North American Aviation (later, Rockwell) was the prime contractor for Apollo and shuttle, and the astronauts felt less connected to those development efforts than to prior ones:

> During the Seventies, an exodus of flight-experienced astronauts accompanied by a rapid increase in the size of the astronaut corps had a net effect of diluting the operational experience in the Astronaut Office. The Astronaut Office had little chance of achieving the operational influence on the space shuttle that we enjoyed during the Mercury, Gemini and Apollo Programs.[86]

Cunningham recounts one unnamed pilot-astronaut, scheduled to fly an upcoming shuttle flight, commenting in 1977 that "for the first time in my career, I look at the most advanced new flying machine and I'm not sure whether I want to fly it or not." Retired Apollo veterans Charlie Duke, Al Worden, and Dick Gordon echoed these sentiments in a press conference in 1979, calling the shuttle, produced on the cheap and never fully meeting expectations, the least safe space vehicle NASA had ever flown.[87]

By 1980, though, astronauts no longer found themselves in positions to make crew decisions or steer hardware design. Deke Slayton, Alan Shepard, and Robert Gilruth had retired, and the autonomy astronauts had earlier enjoyed in crewing flights was quickly replaced by decisions from NASA's civilian management. NASA managers' assent—particularly that of Robert Gilruth and George Mueller—had always been required on crew assignments, but Slayton and Shepard had enjoyed enormous authority and had seldom seen their choices overruled. Following their departure, several astronauts later recounted, George Abbey routinely overruled assignment recommendations from the Astronaut Office under John Young, and reputedly favored certain 1969 group astronauts (like Crippen) over others and sowed dissent within the Astronaut Corps. Abbey, a former air force pilot, had joined NASA in 1967 and assumed increasing duties through the 1970s, 1980s, and 1990s, including, eventually, acting directorship of the Johnson Space Center. An aggressive advocate for the International Space Station and a shrewd political operator, Abbey was admired by many within NASA: Apollo 16 astronaut Thomas K. Mattingly later spoke approvingly of his leadership and the extensive behind-the-scenes influence he carried, but other astronauts were less positive.[88] While Abbey's determination to ensure that all astronauts flew in space at least once earned him grudging respect, many astronauts regarded Abbey as an inscrutable tyrant inclined to play favorites in flight assignments.[89]

Astronauts of the 1960s typically had a sense of where they stood in relation to their peers and usually agreed with Slayton's decisions, but as the first shuttle missions approached, astronauts wondered which mission they would fly and when and found themselves unable to predict the roster as they had done with Apollo.[90] Abbey, nicknamed "Darth Vader" by some astronauts who served under him, jealously guarded the deliberations and articulated no reasons for his decisions, which struck astronauts as less reasoned than the similarly mysterious, but seldom controversial crew picks of Slayton and Shepard.[91] "[N]o one knew exactly how this was going to work," Hartsfield recalled. Hartsfield was "extremely disappointed" not to be selected to be among those to fly one of the ALT missions; "I thought that having developed the flight control system, I'd be in a good position. So did a lot of other people...but we learned along the way...crew assignments are strange things."[92]

Eventually, Abbey assumed many duties formerly assumed by the chief astronaut, John Young, including informing astronaut applicants of their acceptance into NASA and crews of their flight assignments.[93] Instead of receiving word from the Young, who as chief astronaut and mission commander would normally be expected to have made the decision, it was Abbey who casually asked Crippen, while they examined shuttle Orbiter *Enterprise*, whether he wished to join Young in piloting *Columbia* on its maiden voyage. In a 2006 interview, Crippen claimed not to have known why Abbey had chosen him. ("Beats the heck out of me," Crippen explained.)[94] Though expecting an early flight, Crippen had not expected the shuttle's maiden voyage. "It blew my mind," Crippen recalled.[95] Challenging such decisions, fellow MOL transfer Hank Hartsfield realized, did little good. "You just smile and press ahead."[96] Hartsfield learned of his later crew assignment when, summoned mysteriously to Abbey's office, he shared an elevator with several other astronauts, who quickly realized they were heading for the same meeting.[97] Indeed, Gordon Fullerton noted, ignoring the politics or crew selection seemed the only path through it. "There was some shuffling around on who would fly what and all that, all happening at levels above me," Fullerton recalled of the circumstances surrounding his first command assignment. "Somebody would write a book about crew selection some day, maybe. Not me, I want no part of it."[98]

"Orbital Decay"[99]

Throughout the 1960s, the working lives of astronauts remained controlled by a surprisingly small number of individuals, most of them astronauts themselves. But while these individuals were often able to recruit, select, train, and assign their replacements to fly new spacecraft, they could not insure the survival of the programs that built them. As NASA transformed itself from crash program to a large government technical bureaucracy struggling for funding, it replaced its retiring veteran spacemen with new groups of journeymen pilots and scientists less able to mold the space program in their own image. Once heralded as individuals, America's remaining astronauts of the 1970s became dutiful, fungible soldiers submitting to a protracted training regimen and an often oppressive hierarchy in the hope of one day hurtling through the cosmos. Like the late 1960s college students that Charles Reich saw radicalized, seemingly overnight, in *The Greening of America*, NASA veterans like Scott Carpenter, Gordon Cooper, and Wally Schirra eventually cast themselves as free spirits unfairly micromanaged by the bureaucrats of NASA's oppressive "Corporate State."[100]

How such men would respond to this kind of work environment, or what to do with such men when they left it, were questions NASA's 1959 Selection Board had never anticipated. NASA examiners had sought individuals they thought most likely to look good on television and function well during short periods of high stress.[101] While NASA was always aware that its astronauts would be in the public eye, little thought had been given to selecting people psychologically adapted to public scrutiny, protracted training, or the permanent celebrity.[102] Indeed, returning to their selection criteria in 1964, Ruff and Korchin admitted to overemphasizing "stress resistance" in the initial protocols and failing to pay sufficient attention to applicants' "adaptability to complex and demanding social situations."[103] As most of the astronauts had been able to avoid public embarrassment, though, Korchin and Ruff could conclude that, "as every television viewer knows, not only have men performed well as pilots, but they have also risen beautifully to the demands of being public figures."[104] By the 1970s, though, astronauts appeared less able to maintain the charade.

Deke Slayton, after a decade of waiting, finally found a flight on the ASTP mission, but for most of America's first seven astronauts, a career in spaceflight had been a disappointment, with only one reaching the Moon. By the mid-1970s, astronauts had realized that NASA's contracting operations now offered an uncertain path to promotion and few opportunities for advancement. Dick Gordon, awaiting command of Apollo 18, watched helplessly as his mission was cancelled and his crew was broken up; by the early 1970s, he and almost all of the other astronauts of the 1963 selection were retiring.[105] Released from NASA's service after the conclusion of Project Mercury, military personnel were expected to resume their military careers, but having been separated from military work for a decade or more, the astronauts often found their services unwilling to reinstate them at the command level they had expected—astronaut skills weren't respected or transferable to other duties.[106] Wally Schirra eventually retired from the navy when he left NASA, convinced that any command he received would be a public relations position without any real authority.[107] Inertia, fear of failure, and financial woes, Walter Cunningham suspected, kept many astronauts from leaving NASA even earlier, especially among military officers looking to fill out the twenty years of service necessary to retire with benefits.[108] Meanwhile, the consensus among scientist-astronauts was that NASA training had ruined them for other work; most of NASA's scientist-astronauts and all of its MOL transfers chose to remain in NASA despite little assurance they would ever fly.

Once retired, former astronauts increasingly found themselves on the margins of the space race, working as managers and pitchmen for space and aviation-related industries, involving themselves in often shady business dealings or, increasingly, administrative posts at NASA. While some astronauts prospered after leaving NASA (most notably John Glenn and Jack Schmitt, who were elected to the US senate), others, particularly those who had visited the Moon, were visibly changed by the experience. Some veteran astronauts returned from space so unsure as to their future that the only career that seemed right for them was "ex-astronaut," a new kind of celebrity about which the public demonstrated a morbid, almost bottomless curiosity.

Retiring pilots suddenly freed from exclusive media contracts and confidentiality requirements found few restrictions on sharing their disappointments; Mike Collins, in 1974's *Carrying the Fire*, admitted to finding few Apollo-worthy jobs after the Moon, but opened himself to the possibility that he simply wasn't the same man who had left Earth. Unexpressive even before they flew in space, many astronauts returned with their affect flattened, unprepared for normal life. "Not many things seem quite as vital to me any more...it takes a lot more to make me nervous or to make me blow my cool."[109] To Collins, both he and his crewmates Neil Armstrong, Buzz Aldrin (and from Gemini X, John Young) had been transformed—at least as far as he could tell, considering that "I am not as close to any of them as the flight experiences we have shared might indicate."[110] After returning from Apollo 11, though, Collins found some satisfaction steering the Smithsonian Institution's National Air and Space Museum toward the opening of its new facility on the Washington, DC Mall.

Certain other astronauts (most often more junior ones), sought solace in unexpected pursuits or experienced emotional distress upon leaving NASA. Alan Bean, LMP on Apollo 12, pursued painting upon his return from Skylab, while other Apollo astronauts turned to fringe, spiritual, or progressive causes: "[P]rofoundly affected" by his Moon trip, Apollo 14's Ed Mitchell researched extra-sensory perception full-time after his return.[111] Apollo 15's James Irwin established a Christian ministry and embarked upon a quest for Noah's Ark.[112]

Apollo Commanders—all veteran spacemen—tended to respond to lunar flight with the usual quiet aloofness; in press conferences, Apollo 11's Neil Armstrong never broke character from the sincere, humorless test pilot or entertained notions of grandiosity.[113] After his return from the Moon, he avoided the media attention his accomplishment garnered, living as a near-recluse in a "castle surrounded by a moat full of dragons," to the chagrin of some within NASA who hoped he might be a more aggressive spokesman for the space program.[114] Command Module Pilots like Collins, though, were often modest and self-effacing, and insulated themselves from trauma merely by avoiding lunar descent and its associated celebrity.[115]

More often, it was the most junior astronaut, flying (often for the first time) as Lunar Module Pilot, who bore the burdens of interplanetary flight most heavily. Apollo 11's Buzz Aldrin had been extremely conscious of the magnitude of the journey, had lobbied to be the astronaut to step out of the LM first, and as practicing Episcopalian, took communion on the lunar surface. Adrift after his return, Aldrin found solace in alcohol and secretly checked himself into a hospital for what he later described as a "good, old-fashioned, American nervous breakdown."[116] Aldrin, in his 1973 memoir, spoke of a driven boyhood and a "melancholy of all things done," upon his return to Earth, but also wondered whether spaceflight had affected him physically. On their way to the Moon, cosmic rays that had flooded the cabin during Apollo 11's flight; these had produced flashes in his and Armstrong's eyes and, Aldrin feared, destroyed his brain cells.[117] Aldrin had a family history of suicide, but Collins speculated that unchecked ambition and an overbearing father, as well as a certain amount of resentment for not being first to walk on the Moon, were to blame for Aldrin's "incapacitating depression".[118] Even LMPs who had not stepped foot on the Moon, though, occasionally found their perspectives changed: Rusty Schweickart returned to Earth after Apollo 9's Earth orbit flight describing his spacewalk as a metaphysical experience, taking up transcendental meditation and helping to start a telephone crisis hotline in Houston.[119]

To Apollo 7's Walter Cunningham, the "postflight depression" that followed spaceflight was "professional not personal." "We had completed the mission, a goal for which we had trained and dedicated and sacrificed most of our lives, but there was no way to know if we'd have another."[120] Irwin's new interest in spiritual matters puzzled crewmate Al Worden, who at first doubted Irwin's sincerity. Apollo astronauts were routinely questioned about their spirituality upon their return to Earth, and, "Jim's response seemed to answer the constant inquiry a little too neatly for me." Worden speculated that the collapse of his flying career at NASA drove Irwin to find fulfillment in something other than the space program.[121] In fact, spaceflight hadn't so much changed them men as given them license to reveal aspects of their personalities that they had formerly shielded from public view. Few among them expressed interest in subjects that hadn't consumed them in some way before their flights. Expecting spaceflight to change them was, to Cunningham, the biggest disappointment. In a 1972 *New York Times* interview, Cunningham offered his own theory of Aldrin's difficulties: "Buzz Aldrin doesn't have anything to say. . . .He's an example of a guy who wanted to have a transcendent experience and was disappointed because he didn't have one."[122]

Journalists of the 1970s were particularly sensitive to suggestions that space flight had worn heavily on the astronauts' psyches, but the evidence of such transformation was anecdotal at best. NASA, astronaut Jim Lovell later recounted, routinely assigned a psychiatrist to recover crews to assess whether returning astronauts were "spaced out or raptured to death," but found no such evidence.[123] Yet, magazine and newspaper articles of the 1970s surveying NASA's retired astronauts often found them wistful and transformed.[124] Such accounts, coming at a time of increased introspection for all Americans, satisfied reporters who had long wondered why astronauts had not reacted to spaceflight with a great sense of wonder. Some astronauts, to be sure, had returned unchanged, but "in the behavior of others," Howard Muson wrote for the *New York Times* in 1972, "there is more than a touch of the eccentric, and a large dose of trouble, almost a mythological element: the wandering hero back among his tribe, after stealing the sacred fire and grappling with terrifying demons, condemned to ask tough questions."[125] Journalists had long wondered why astronauts had not gushed in emotion at their travels. "To our relief," Muson wrote, "they now seem much like us."[126]

Meanwhile, popular representations of astronauts offered spacemen more introspective about their adventures and more open to progressive ideas about humanity, portraying the astronauts as men who, in their cosmic loneliness, could realize truths hidden to terrestrial men. Astronauts often fueled such representations. One of the earliest fictional spacemen to suggest such a transformation (and complicate the image of the American spaceman) was English singer-songwriter David Bowie's "Major Tom," described in his 1969 single "Space Oddity." At first, Major Tom appears in the song as cut from the same cloth as the Original Seven, stalwart yet personable. Ground controllers dispatch him with the usual platitudes ("may God's love be with you") and journalists pepper him with inane questions ("the papers want to know whose shirts you wear"). Before too long, though, Tom loses his bearings, visceral sensation replaces careful observation, and he is consumed by nihilism ("sitting in a tin can...there's nothing I can do"). To be sure, this new conception of the astronaut was not immediately popular. "Space Oddity," though popular in Britain, failed to chart in the United States until its rerelease in 1975. By then, the idea of the impaired spaceman, fueled by press accounts of returning Apollo crews, was part of the American cultural vocabulary. By 1976, the idea of the returning astronauts as "basket cases," was so common in popular culture that Rip Torn's cynical university professor in *The Man Who Fell To Earth* could declare the point casually (despite a cameo appearance by the retired but very sane astronaut Jim Lovell), as if the premise were common knowledge.

In 1973's film adaptation of William Peter Blatty's novel *The Exorcist*, a young girl intrudes upon a cocktail party in her movie star mother's posh Georgetown home to tell one guest, a NASA astronaut, what only a child possessed by the devil would dare say to his face—that he is "gonna die up there."[127] Blatty's morose astronaut from *Exorcist* eventually returned to the screen, consumed by madness in his 1980 film *The Ninth Configuration*.[128] Where the spaceman, played by Richard Callinan in *Exorcist*, is a tall, gregarious, square-jawed role model, Scott Wilson's character in *Ninth Configuration* is a slender poet confined to a creepy psychiatric hospital. Having aborting his lunar flight just before launch, Marine Corps captain Billy Cutshaw is seen in flashback dragged from the rocket in his spacesuit, screaming that "nothing's up there." The stress of repeated, lonely orbits around the Earth have apparently broken the man; he refuses lunar duty, consumed by the fear that he will (as the possessed girl warned) die, alone, in space, "so far from home," with no God to comfort him. In the film's most famous and perplexing scene, Cutshaw's psychiatrist imagines a space-suited

Cutshaw on a lifeless moonscape encountering a crucified Jesus, a possible reference to James Irwin, who claimed that spaceflight had made God more apparent to him.[129]

Shortly after the crew of Apollo 16 returned on May 19, 1972, Elton John's LP *Honky Château* debuted with a memorable track about a weary "Rocket Man" trying to reconcile himself to the petty discomforts of his profession without completely losing his mind. While it was a 1951 Ray Bradbury short story, and not NASA's own spacemen, that inspired songwriting partner Bernie Taupin's lyrics about a Mars-bound, clock-punching spaceman, certain aspects of the characterization seemed to ring true.[130] Routine, not adventure, defines the Bernie Taupin's Rocket Man; spaceflight is "just a job," one whose technical specifics are barely within his comprehension. As his beloved wife and children wait for his return, the Rocket Man steels himself to the possibility of living—and dying—alone in space, and yet he cannot quite bring himself to walk away from his vocation. Listeners are left to wonder why; the Rocket Man, like many of his real-life counterparts, has no way with words. All the spaceman can articulate is the idea that spaceflight somehow liberates him: "I'm not the man they think I am at home," and for all his dissatisfaction, he cannot easily walk away from an identity he enjoys only when hurtling through the cosmos.

The unnamed protagonist of "Rocket Man" is neither scholar, nor soldier, nor scientist: he is a five-day-a-week hired hand, describing himself with a kind of subtle, insidious despair that labor sociologists of the 1950s found in dissatisfied white-collar workers. Even in the 1950s, writers had looked to the day when space piloting would join the pantheon of ordinary professions: the 1955 film *Conquest of Space*, based on the writings of Willey Ley, focused on the experience of an international crew of technical experts on the first flight to Mars. Anticipating an era when spaceflight would become routine, the film suggested that crew members would complain endlessly about the discomforts of their voyage and perhaps even be drafted unwillingly into space service due to their skills and abilities.[131]

Of all of the depictions of astronauts in the 1970s, the protagonist of "Rocket Man" was, perhaps, closest to the truth. In *The Man in the Gray Flannel Suit*, the classic postwar novel of the new "company man," a war-damaged business executive trades his emotional stability for material success as a corporate cog.[132] The new Space Age, *Time* wrote as early as 1957, had already seen a proliferation of such men: engineer-soldier-academics who could be anyone or no one—men in "gray flannel" suits who lingered in block houses and control rooms organizing America's ballistic missile program.[133] By 1975, most Apollo veterans, faced with years in training and few flights on the horizon, had chosen chose to leave NASA and confront the dislocation and confusion of life after space travel. Meanwhile, NASA's MOL transfers and scientist-astronauts steeled themselves to the possibility of extended training, tedious administrative duties, thankless engineering work, and diminished authority to continue in their chosen profession. The man in the gray flannel spacesuit had finally arrived.[134]

Chapter 5

Public Space

In January 1972, an unexpectedly enormous group of science fiction enthusiasts descended upon the Statler-Hilton Hotel ballroom in New York City to share reminiscences of a television series whose brief run on NBC had ended in 1969, the same year Apollo astronauts first stepped on the Moon.[1] The series, set in Earth's distant but recognizable future, chronicled a thinly disguised version of the National Aeronautics and Space Administration (NASA) at the peak of its 1960s influence—an organization of outsized personalities serving in a progressive, quasi-military organization devoted to exploration of the galaxy.[2] Three hundred years into the future, President Lyndon Johnson's Great Society had been achieved in full—on a planet without poverty or social strife, Americans (who appeared to dominate Earth's future) seamlessly integrated their terrestrial activities with flight into the deepest reaches of the galaxy. Space travel had brought opportunity and prosperity, including new sources of energy and wealth, colonies and allies, and knowledge exchange with extraterrestrials. Ironically, *Star Trek* proved even more popular after the Apollo program that had inspired it had waned: its vision progress and internationalism remained deeply attractive to many Americans. NASA, aware of their enthusiasm, answered in kind. Responding to a request from the *Star Trek* convention's organizer for the loan of Moon rocks to celebrate the event, NASA offered a trailer's worth of hardware and memorabilia.[3]

NASA's creation in 1958 had been intended, in part, to ameliorate public dissatisfaction with America's disorganized space exploration efforts. Throughout the 1960s, NASA managers had attempted to curry favor with the American public, with astronauts often dutifully fulfilling their role as space-age public relations specialists. Their successes helped inspire a generation of science fiction about spaceflight, and a nation of space fans whose conception of space exploration at first mirrored, and then increasingly diverged from, the space program that had inspired it. Rather than ignoring such voices, NASA, throughout the 1970s, cultivated support among science fiction fans, altering its funding strategies to harness and, if possible, manipulate public fascination with space travel, and making its astronauts into new popular icons for a new decade. NASA would, with varying degrees of success, anchor the space program to a variety of popular causes—environmentalism, internationalism, civil rights—developing the space shuttle to accomplish all of them. These efforts would have consequences for decades to come.

Constructing the Astronaut

At the inception of NASA's human spaceflight program, agency leaders were torn between their desire to minimize the celebrity of NASA's astronauts and their need to harness the pilots' notoriety to sell the space program to the American people. It did not take long, though, for the astronauts to recognize that they could exploit the public's fascination with spaceflight to further their goals in the program. From the earliest days of Project Mercury, NASA's astronauts leveraged their high public visibility to ensure that human piloting remained central to Project Mercury. When Gordon Cooper casually remarked to reporters that NASA had not provided them with jet planes with which to maintain their flying proficiency, for example, a public outcry resulted in the transfer of several supersonic jets from the air force, as well as consternation among NASA managers that an astronaut had so easily circumvented them by appealing to reporters.[4] Even the language that engineers used underwent revision in response to astronaut preference and public perception; Mercury "capsules" were eventually renamed "spacecraft," which the astronauts would fly as "pilots."[5] "Capsules are swallowed," wrote Mike Collins, "one flies a spacecraft."[6]

Military pilots had long enjoyed the privilege of naming their aircraft and designing its nose art; once in NASA, the astronauts also asserted this authority. Initial efforts by the astronauts to personalize their vehicles were not greeted warmly; as Gordon Cooper later wrote, NASA administrator James Webb preferred numbered flights and, seemingly, numbered crew members. In addition to NASA's numerical designation schemes, Mercury spacecraft ultimately also carried an informal name chosen by their crews, referred to in media reports and painted on the vehicle itself. The names proved to be good public relations, and in 1965, Webb authorized the astronauts to create mission patches for each flight, designs that would be sewn onto their spacesuits and would contain the names of the crew and also symbolic artwork representing the mission's activities. The patches, designed for public as well as private use, often colorfully referenced popular culture, from Conestoga wagons to popular musicals.[7]

Efforts to enlist public support in the astronauts' work accelerated with Project Apollo, with mixed success. While NASA hoped to portray the men as peaceful explorers, public audiences were often fascinated more with the military aspects of space work. Popular depictions of astronauts in the 1960s emphasized the astronauts' military backgrounds, courage, and unique fortitude of character. Stoic, fictional astronauts made frequent appearances on Rod Serling's innovative, surrealistic fantasy series, *The Twilight Zone*

(1959–65), bravely confronting new worlds, strange creatures, and death.[8] In "Death Ship," one noteworthy episode from 1963,[9] the protagonists refuse to acknowledge mounting evidence that they've already perished, and determine to keep flying.[10] The theme of the astronaut as warrior was a common one from the 1960s through the 1980s; motion picture depictions of astronauts often assigned to them military skills that no real astronaut was expected to possess. The astronauts in the James Bond installment *You Only Live Twice* (1967) fight their captors with expert hand-to-hand combat;[11] later, in 1979's *Moonraker*, shuttle astronauts are laser-armed space warriors who help capture an enemy orbital outpost.[12]

NASA did not always approve of such characterizations. It often withheld permission from filmmakers seeking to use its name or logo in fictional programming, and, where possible, attempted to steer popular representations of astronauts to suit its political agenda. While the Moon landing program eventually grew to embrace a variety of humanistic and spiritual values, NASA, in internal documents, conceded that national security goals were paramount: to "prove American technological superiority without military confrontation, to build a new level of national pride and prestige, and to create a base of science and technology for the future."[13] Yet, Kennedy administration officials and NASA handlers tirelessly promoted the image of the astronauts as good-natured civilian explorers in government service.[14]

Larry Hagman's fictional character "Major Nelson," from the television comedy *I Dream of Jeannie* (1965–70), is a capable air force officer who manages to do his duty despite good-natured interference from Jeannie, a temptress he found in a bottle on a beach after returning from space. Hagman's spartan space traveler is the show's role model, intended to be "one of NASA's best."[15] NASA vetted scripts for the show and collaborated with associate producer Sidney Sheldon to ensure accuracy. (Nelson's colleague, the fictional Major Roger Healy, played by Bill Daily, is, by contrast, a pompous extrovert reputedly based on astronaut Alan Shepard.[16]) In one letter to Sheldon, NASA objected to Sheldon's emphasis on depicted astronauts as military personnel, nearly always in uniform. To NASA Public Affairs officer Walter Whitaker, the media's obsession with the military aspects of the human spaceflight program was a problem to be rectified by deemphasizing military plotlines in story scripts. Chastising the producers of *Jeannie*, Whitaker cautioned, "[we] try project the image of the program as peaceful, scientific exploration of space. This is an important part of our international relations."[17]

The dangers to which the astronauts exposed themselves in the service of the country, the image of supportive wives and children, and the specter of death resonated with audiences, as did comments that distanced the astronauts from the godless communism of their Soviet counterparts.[18] Though aware of many of the astronauts' extramarital dalliances, reporters chose to suppress the reports: the public preferred to see NASA's astronauts, like the dour Major Nelson, as men demonstrating an almost monastic devotion to spaceflight. Like any group of military officers in wartime,[19] the astronauts were males tasked with manly responsibilities, supported, wrote *Life* magazine in 1959, by "brave wives" far away who raised the children and tolerated their husbands' risk-taking.[20] Astronauts who pushed the envelope of conduct could receive private reprimands or worse, but to a reverent public unaware of the astronauts' dalliances and NASA's discipline, astronauts appeared as icons of self-control, resolutely heterosexual but romantically aloof.[21]

While NASA managers and the astronauts attempted to control public sentiment, both faced a public with its own ideas about who, exactly, the spacemen were. By the early 1970s, astronauts constructed as humorless Cold Warriors no longer held the same appeal to the American public. With the budget for human spaceflight contracting in the late 1960s and early 1970s,[22] a growing counterculture increasingly demanded a space program more reflective of popular aspirations: racial and gender integration, environmental awareness, planetary exploration, space settlement—or none at all.[23]

From Skylab to Space Colonization

In NASA's effort to sell spaceflight to a new generation of Americans, astronauts would find their next great role: as unlikely spokesman for environmentalism. As the Apollo 11 astronauts returned to Earth, Mission Control had radioed to them the news of the week, including the nation's number one song, "In the Year 2525," by Zager and Evans, a mournful prediction of humanity's extinction on a wasted and exhausted Earth.[24] With their planet in tatters, many argued at the time, the economic costs of space travel outweighed any potential benefits. "Our astronauts glide through space at 25,000 m.p.h.," physicist and activist Ralph Lapp wrote in February 1969 as he assessed spaceflight's future, "while our streets are choked with bumper-to-bumper traffic. We shoot ultra-sophisticated probes to sample the atmosphere of Mars and Venus while our cities are enveloped in polluted air. Space men nibble on expensively developed diets while Biafran children die from malnutrition."[25] Concern about the future of the Earth, while at first threatening the space program, ultimately provided NASA with a new rationale for it, transforming its spacemen from soldiers into environmental sentinels and activists. Astronauts inclined to say more in public would find the public ready to hear their words. The image of the astronaut as hippie, like that of the astronaut as soldier, was largely fabricated by the popular writers, but proved irresistible to the public and useful to NASA managers.

The "new" American environmentalism of the late 1960s and early 1970s had deep connections to the space program; its ideas drew heavily from "human factors" spaceflight research of the 1950s and often employed spaceflight vocabulary. Human factors research had theorized space vehicles as self-contained habitats recreating terrestrial ecosystems, suggesting that the role of the space vehicle engineer (like that of the submarine designer or fallout shelter builder) was to understand Earth ecology so that one might reproduce it in a closed container. This work occupied a number of biologists and ecologists who, while minimally influential in the space program of the 1960s, eventually applied their research with more success to environmental studies.[26]

The extrapolation of this "cabin ecology" movement to environmentalism was relatively rapid; historians generally credit the first complete discussion of the term "Spaceship Earth" and its implications to architect Buckminster Fuller some time around 1963, though the term first appeared in a children's book and the first reference to the planet Earth as a vessel moving through space actually appeared in 1879. The idea of a planet as a vehicle drove home the idea of Earth as a self-contained vessel in which humans had to cooperate or risk death by resource depletion. In 1969's *Operating Manual for Spaceship Earth*, Fuller even announced that as beings flying through space on a self-contained life support system, "we are all astronauts."[27] The "Spaceship Earth" concept would become one of the more memorable icons of the environmental movement.[28]

Meanwhile, throughout the late 1960s and early 1970s, NASA had slowly become a leading environmental organization, though in ways that were often not readily apparent to the public. NASA's exploration mission positioned it as both a major polluter and a major contributor to pollution research; manufacturing, assembly and flight operations of rocket vehicles consumed huge tracts of land, some of which was contaminated by nuclear and chemical residues. NASA's near monopoly on space-based remote-sensing equipment also gave NASA a privileged position to record data about Earth's atmosphere, landmasses, and oceans. This information was vital to its space going endeavors: threat of pollution from the solid propellant rocket motors engendered scientific studies that placed NASA at the forefront of pollution studies research. In 1971, public controversy, supported by NASA research, helped terminate a decade-long effort to build an American supersonic passenger jet, whose high cost, deafening noise and ozone-destroying exhaust made it unpalatable to a growing, vocal group of environmentalists.[29]

More significantly, NASA had assumed responsibility for planetary protection from dangerous extraterrestrial organisms, both demonstrating the potential environmental dangers of space exploration and highlighting NASA's increasing sensitivity to such concerns. While theorists heralded the day when spaceflight would rescue humanity form the Earth's eventual destruction, NASA's space biologists feared that in the near-term, astronauts would return from the Moon tracking space dirt teaming with dangerous microbes. Michael Crichton's *The Andromeda Strain*, published only two months before the flight of Apollo 11 and adapted for film in 1971, depicted scientists struggling to control a runaway space microbe inadvertently released on Earth. Collected by a space probe for military applications, the germ weapon turns on its discoverers, clotting blood and eating plastic until scientists eventually bring it under control. Despite the seeming outlandishness of the plot, NASA feared that such a scenario was plausible enough to quarantine early Apollo crews upon their return to Earth.[30] "It's possible that a biologically active plant spore from the moon could destroy all plant life on earth,"[31] the *New York Times* quoted one space biologist, months before *Andromeda Strain* hit bookstore shelves.

Of these efforts, NASA's ability to actually view the Earth from afar proved most significant. Photographs of Earth as a life-giving body in the sterile vastness of outer space gave environmentalists new images around which to organize themselves, and made heroes of the men who took the pictures.[32] In 1966, counterculturist Stewart Brand, a pivotal figure in many San Francisco area progressive movements, led a campaign for NASA to release imagery of the whole Earth from space to inspire environmental awareness and positive action, a goal Brand supposedly incubated while under the influence of psychedelic pharmacopoeia. "How can I make this happen?" Brand wondered, tripping on LSD on a San Francisco rooftop, "it will change everything if we have this photograph looking at the earth from space."[33] Contrary to popular belief, though, no whole Earth photograph existed in 1966, let alone one taken by astronauts, who had not yet ventured far enough from their home planet to view more than a glimpse of Earth's curvature.[34] In fact, such wide-angle photography was not a priority of either civilian or military space programs. While various American and Soviet satellites took a series of long-distance images of Earth in 1966 and 1967 (including NASA's Lunar Orbiter and the air force's DODGE spacecraft), NASA's ATS-3 satellite, launched in November 1967, transmitted the first high-quality color image, and it was this photograph that graced the cover of the first *Whole Earth Catalog*, published in fall 1968 by the California-based Portola Institute under Brand's leadership.

For astronauts who struggled for words to describe their experiences in space, statements acknowledging Earth's beauty had always been among the easiest to make and most readily absorbed by the press and public. Sometimes, astronauts of the late 1960s even returned to Earth with what appeared to be genuine concern for Earth's habitat. Boisterous Apollo 7 commander Wally Schirra was so incensed by the smog covering California that he sent photographs of it to Governor Ronald Reagan upon his return, a presumptuous act for an astronaut at the time.[35] For a brief period in the 1970s, NASA wondered whether environmentalism might even become a new theme around which to organize its efforts, one that would turn its explorations into works of advocacy and its astronauts into public intellectuals.

Many in the environmental movement credited the astronauts with galvanizing popular enthusiasm for Earth-awareness; their photographs of Earth were even more compelling than those of distant worlds. Coasting around the Moon on December 1968, Apollo 8 astronauts Frank Borman, Jim Lovell, and Bill Anders had conveyed their blessings to everyone "on the good Earth" and snapped photographs of a crescent Earth rising above a desolate lunar horizon against an inky black sky.[36] One of these images was reproduced as a 1969 stamp and became an even more iconic a fixture of the nascent environmental movement, described, in 1971's *The Last Whole Earth Catalog*, as the photograph "that established our planetary facthood and beauty and rareness...and began to bend human consciousness."[37]

Anders would later note that while the Apollo 8 astronauts had come "all this way to explore the moon…the most important thing is that we discovered the Earth."[38] Even the stolid Borman was moved to announce that protection of Earth's atmosphere was "immediate" priority, "regardless of the economic considerations."[39]

Stirring images and poetic commentary like this eventually inspired new conceptions of the astronaut as countercultural figure able, by virtue of his unique journey, to "tell it like it is." In the products of environmentally-aware art collectives like Ant Farm, the astronaut soon became a symbol of physical and psychic rebellion, clad in minimalist white coveralls, surrounded by pop-art renderings of the American flag, and defiantly gripping a portable video camera.[40] As environmental spokesmen, though, astronauts were imperfect specimens. What inspired the Apollo 8 crew was less a call to environmental action than Earth's colorful beauty amidst a vast dark expanse. Apollo 8 astronaut Frank Borman had been the first to see the famous Earthrise and he photographed it with black-and-white film. The mission plan had not included photography of the Earth from lunar orbit, Anders recalled, and he was at first annoyed to find Borman wasting film on a target not in the mission plan. (When all three crew members saw the sight, though, they began furiously taking pictures.) In the years following Apollo 8, the astronauts later claimed to see "all sorts of things" —ozone depletion, global warming, resource use—in the striking image.[41] Apollo 8's crewmembers, though, have spent as much time praising the image as arguing over which of them had taken it, determined to compete even in the production of hippie art. (Lovell, in the film *In the Shadow of the Moon*, grudgingly gave credit to Bill Anders for taking the famous Earthrise photo, while still hinting otherwise.)[42]

Other inadvertent public outreach efforts by the NASA and the astronauts were similarly successful. On NASA's final lunar landing flight in December, Apollo 17's crew turned one of its 70 mm Hasselblad cameras toward the receding Earth, taking what would prove to be the most famous photograph of that body ever, and the most reproduced image in history. "The Blue Marble," a photograph of the whole Earth unobstructed by shadow taken by the Apollo 17 crew at a distance of 18,000 miles, was distributed to the press soon after splash down and became a fixture of the environmental movement.[43]

Though it had not intended to foster it, NASA briefly embraced and encouraged such environmentalism. In early 1972, NASA deputy administrator George Low had solicited ideas from its Space Program Advisory Council on "what the space program should be and is likely to be in the 1980's [sic]." Development of a human spaceflight capability, epitomized by Apollo, had been the "key project" of the 1960s, but Apollo was approaching its end, with the first flight of a replacement vehicle years away.[44] To some, though, NASA's identification with transitory, high-profile projects was counterproductive; the agency needed to organize itself instead around an enduring theme that would resonate with the public, and none was better than environmentalism. By 1972, NASA had recognized the immense popular enthusiasm generated by its Earth photography, and its position on the forefront of environmental research. Redirection of NASA's efforts from Apollo to less expensive Earth studies suited its shrinking budgets, and NASA adjusted to the new political climate.

"Earth Resources Technology Satellite 1" (later dubbed "Landsat"), launched on July 23, epitomized this new mission for NASA. The first in a series of observation probes intended to make Earth resources imagery available to the wider public, Landsat was the antithesis of Apollo's almost martial organization. From its inception, space historian Pamela Mack has written, Landsat had no clear goal; rather, NASA hoped, the mere availability of visual and infrared photography of the Earth's surface, administered by a public/private consortium, would enrich a diverse assortment of interest groups concerned about land use and natural resource management.[45]

As one 1973 memo from NASA assistant administrator for Public Affairs John Donnelly to administrator James Fletcher noted:

> The more I think about it, the more convinced I become that lacking a "driver" of the Apollo type, we've got to arrive at an agency rationale and get away from the project-oriented mode, wherein we spend time, money and effort emphasizing projects that quickly become obsolete and which effort creates, albeit unintentionally, a fractionated non-cohesive public image of the agency among the non-cognoscenti.[46]

Instead, Donnelly wrote, NASA should describe its next generation of space vehicles in terms of their potential environmental benefits; doing so would provide NASA with a unified message that would transcend particular program and tap into a wellspring of existing popular enthusiasm:

> In summary, to repeat our reasons for recommending the environmental approach:
>
> 1. It defines a purpose and direction for the agency.
> 2. It encompasses the activities of both manned and unmanned activities.
>
> 2a. It provides a cohesiveness in terms of public image that is now lacking.
>
> 3. It attaches us to a movement that ranks high on the list of public and congressional priorities and makes us "relevant."
> 4. It's easier to be "for" than "against."[47]

Fletcher received the memo positively; writing to Low that it required serious discussion.[48]

Never before had NASA found itself so close to embracing the counterculture that had opposed it. The hope of many in the environmental movement was that NASA might deploy its seemingly infinite technological arsenal to solve the environmental crisis. Subtitled "Access to Tools" and intended as a guide book for do-it-yourself technologies available by mail order, Brand's *Whole Earth Catalog* not only served as a publication outlet for NASA's spaceflight imagery, but subtly praised NASA's work as necessary first steps to a larger purpose: the creation of ecological utopias in space. The *Catalog*'s tools for living contemplated the deployment of high technology in the service of sustainable living, but Brand's philosophy contemplated no limit to the size of the technologies employed. He later advocated the colonization of space as a further evolution of this ideal, hoping to create in orbit the ecological paradise geopolitics would appear to prevent on Earth.

NASA was a significant player in post-Apollo discussions of space settlement; summer studies by NASA's Ames Research Center during the 1970s produced numerous designs and most of the now-iconic images of the proposed space colonies: huge cylinders, wheels, or spheres encompassing Earth-like habitats complete with lush fields, rivers, and populations of ten thousand people, living, working, and socializing amid California modern architecture (Figure 5.1).[49] NASA also financially supported a conference on space colonization at Princeton University led by physicist and space colonization enthusiast Gerard O'Neill, where O'Neill spoke of NASA's space shuttle, then under development, as the vehicle that might assemble the first space colonies.[50]

NASA's increasingly vocal astronauts slipped almost effortlessly into these discussions, suggesting a larger cultural relevance for their work than NASA had anticipated. In popular films of the 1970s, astronauts increasingly found themselves portrayed as mystics with a sentimental attachment to the natural world now under threat. One of the most extreme examples of this new construction of the astronaut was Bruce Dern's portrayal of space botanist Freeman Lowell in the 1972 film *Silent Running*. While ferrying the last

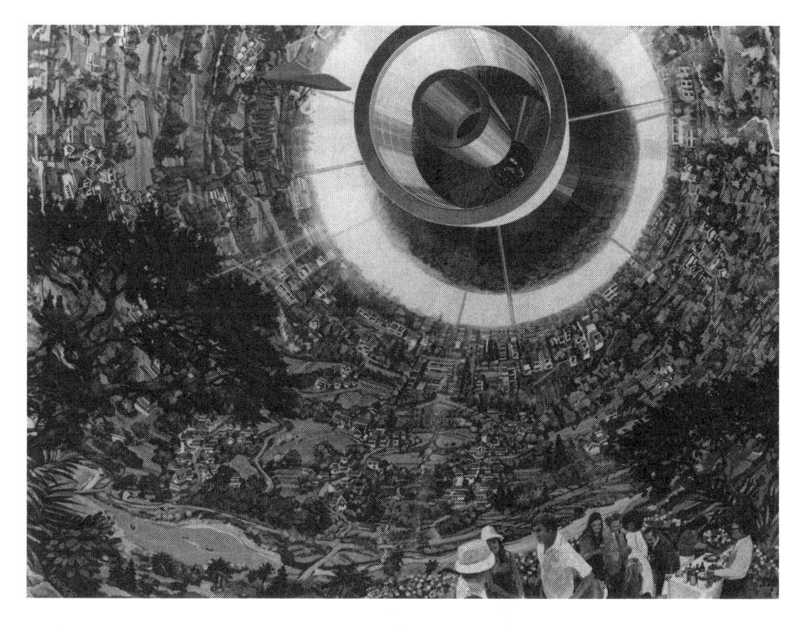

Figure 5.1 This 1970s-vintage NASA artist's conception depicts one design for a future space colony that would serve as a self-supporting home for thousands of people (NASA photo)

remnants of Earth's forests aboard the American Airlines space freighter *Valley Forge*, Lowell cultivates a murderous resentment of his crewmates over their indifference to Earth's botanical richness and mistreatment of the ship's robots. Indeed, Lowell serves in the film as a thinly veiled version of nineteenth-century Massachusetts author, naturalist, transcendentalist, abolitionist, and political dissenter Henry David Thoreau.[51]

Ironically, a similarly organic conception of the space traveler appeared in the Soviet Union at nearly the same time, reveling not in the false nostalgia of American pastoralism, but in the literary traditions of the nineteenth-century Russian intelligentsia. In Andrei Tarkovsky's brooding adaptation of Stanislaw Lem's science fiction novel, *Solyaris*, space authorities send morose psychologist Kris Kelvin to a distant planet to diagnose the odd behavior of a group of researchers. There, he encounters an alien intelligence able to sense human desire and impose thoughts and sensations at will. For Kelvin, the closest thing to heaven that this Soviet hero can imagine is to be alone with his thoughts at his dacha in the Russian countryside, and the alien intelligence eventually indulges him.[52]

Real astronauts occasionally became actors in the space colonization debate, sometimes unwittingly through their photography, or purposively, through public activism. Like several of his colleagues, Apollo 9 astronaut Russell "Rusty" Schweickart—"the closest thing there is to a freak astronaut" one countercultural radio station manager told the *New York Times*, admiringly—was an enthusiastic proponent of both environmental awareness and space colonization.[53] "Many of us, on returning home from space," Schweickart wrote to space colonization enthusiast Gerard O'Neill, "brought back the perspective of a lonely and beautiful planet crying out for a more responsible attitude from its most prolific partner." Environmentalists eagerly embraced such testimonials, but debate on the potential and practicality of space colonies split the nascent environmental movement in two, with space enthusiasts like Brand and O'Neill praising space colonies and former 1960s radicals from Ken Kesey to Louis Mumford finding in them only "infantile fantasies" and a chance to reproduce, in space, the imperialism, militarism, and wasteful excesses of American industrial capitalism.[54]

Upon their return to Earth in 1974 from the longest American spaceflight to date, Skylab 4 astronauts Gerald Carr, William Pogue, and Ed Gibson were the guests of honor at a Congressional hearing that included detailed questions as to the astronauts' views regarding Earth's environmental problems and their solutions. While most early spacemen might have graciously demurred to air such comments in such a forum, the Skylab 4 astronauts eagerly responded, stressing the relative paucity of habitable areas on Earth's surface "we don't have a whole lot of places to live comfortably on this Earth and we must take care of it," Carr noted.[55] Gibson, in a later interview, spoke reverently of the "*fragility*" of the Earth and of how its lack of discernible political boundaries made him a peace advocate. "Why should the people on this plot of ground be shooting at the people on that plot of ground? It's so obvious that there aren't that many differences, that it's all the same earth."[56] Skylab's experimental payload had included Earth observation equipment, but the research was never a priority. Upon their return, though, the mission had suddenly made NASA's astronauts experts on human habitation of space and the solutions to Earth's environmental problems.

By 1974, though, NASA had already moved as close as it ever would to the dream of environmentalism and space settlement; the massive funds required for space colonization were simply unavailable.[57] Vice President Spiro Agnew had lobbied publicly for a piloted Mars mission even before Apollo 11 had lifted off; Nixon silenced him soon after in a 15-minute closed-door session.[58] Wernher von Braun, too, had hoped that the Apollo lunar landings would be a springboard to a mission to Mars—his ambition ever since coming to the United States—but NASA was uninterested, and the agency sidelined him in 1970 to a headquarters job in Washington, DC, far from NASA operations. In June of 1972, von Braun left NASA for work in private industry, succumbing to cancer five years later.

By then, a certain cognitive dissonance about the space program had already begun to set in at the Manned Spacecraft Center in Houston. In one 1972 *New York Times* interview, Chris Kraft speculated generally on extraterrestrial life and NASA's plans for interplanetary spaceflight, even though, as he spoke, NASA was eliminating its facilities devoted to such research.[59] The Mars flight, in particular, had anticipated development of a new propulsion system employing atomic power; NASA had expected it would need nuclear upper stages for Apollo but had developed a chemical rockets sufficient for translunar flight. On December 26, 1972, NASA ended its nuclear research program at its Plum Brook Station in Sandusky, Ohio, with a handwritten memo from NASA Headquarters to Bruce Lundin, director of NASA's Lewis Research Center. Part of the NASA's Lewis Research Center, Plum Brook had been entirely successful as a nuclear laboratory but, after 1961, never again figured prominently in NASA's short-range planning. In the wake of Apollo 17, with no Mars-bound spaceship likely anytime soon, NASA decided to cut its losses despite Plum Brook's accomplishments and respectable safety record. (It did not help that, in an era marked by increasing sensitivity to the dangers of radiation pollution, NASA's nuclear reactors were a growing public relations problem.) Announcing the closure of Plum Brook on January 5, 1973, Lundin deviated from his prepared remarks and blamed NASA for mortgaging the nation's future in deep space exploration and, presumably space colonization.[60] While NASA would attempt to harness popular support for environmentalism in the years that followed, a larger shift in the mission and focus of the agency never materialized, nor did the Congressional appropriations that might have made such a program possible. As Kim McQuaid writes, throughout the 1970s NASA moved fitfully through a series of modest environmental programs, receiving criticism on both the left and the right, and ultimately abandoned its efforts.[61]

International Dreams

While NASA's environment ambitions fizzled, new opportunities arose to use spaceflight as a vehicle for peacemaking, particularly with the Soviet Union.

A key aspect of the *Star Trek* universe had been its faux internationalism: though an American (played by Canadian actor William Shatner) sat at the helm of the fictional USS *Enterprise*, his crew included an African, a Scot, a Russian, a Japanese American, and even an extraterrestrial. To a public increasingly weary of fruitless superpower competition in the late 1960s, *Star Trek*'s vision of a future Earth free of geopolitical disputes resonated well with audiences and with NASA. Unlike space colonies, international cooperation would be relatively economical, especially if it prevented needless duplication of space vehicles in different countries. Soon after announcing his goal of piloted lunar landing in 1961, President Kennedy suggested that a joint lunar mission with the Soviet Union might be preferable, especially if the costs of the flight could be split between the two nations. Internationalizing Apollo, though, was a nonstarter: Soviet authorities were hostile to the suggestion and, in any event, Kennedy had sold the plan to Congress as an opportunity to demonstrate American preeminence in space.[62]

Apollo's success, and the arrival of détente between the United States and Soviet Union under the Nixon administration, created, by 1972, a geopolitical environment more amenable to joint missions. Among the arguments Skylab's defenders had deployed after the success of Apollo 11 was that the space station, unlike further lunar flights, offered fewer public relations opportunities for international cooperation. In a letter to Edward E. David, President Richard Nixon's science advisor, NASA deputy administrator George Low conceded that Apollo was a dead-end for media coverage: "no major new opportunities for international leadership and prestige would likely accrue; and the potential of Apollo for international cooperation is limited." NASA's funds, he argued, would be better spent on Skylab, a program on which NASA had made a "considerable investment to date. . . ."[63]

Skylab, though, would not, in its final incarnation, produce the hoped-for international cooperation. One obvious option—to internationalize Skylab by opening it to Soviet spacecraft—was not pursued.[64] Even if American and Soviet spacemen could not share a space station, they, could, still, rendezvous in orbit, and possibly dock their Apollo and Soyuz spacecraft. If not for the popular literature promoting such a flight, though, the Apollo-Soyuz Test Project (ASTP) would likely never have occurred. In 1964, a novel by Martin Caidin described the hypothetical rescue of an American astronaut stranded in orbit by a second American craft and, more dramatically, by a Soviet spacecraft.[65] Adapted for the screen (with NASA's encouragement) in early 1969, *Marooned* presented the American public with a Soviet cosmonaut sharing with his American counterparts a fundamental desire for exploration and concern for the preservation of human life. The film, which starred Gregory Peck and Gene Hackman among other well-known actors, was filled with heroic fictional characters bearing an eerie physical and emotional similarity to actual spacemen, including Deke Slayton, cosmonaut Aleksey Leonov, and the entire crew of Apollo 7. (One American astronaut was even named "Buzz" in the film, lending the plot an aura of historical fidelity.) Most importantly, the film suggested that in space the

United States and Soviet Union might achieve a kind of accommodation they could not on Earth.

Before traveling to the Soviet Union in 1970 for negotiations on joint American-Soviet space operations, Philip Handler, president of the National Academy of Sciences (NAS), had enjoyed a special screening of *Marooned*, from which he emerged enthusiastic. The initial response of his counterparts at the Soviet Academy of Sciences on the subject of joint flights had, until 1970, been apathetic. During the Moscow negotiations in May, though, Hander deviated from the usual fruitless entreaties to describe the plot of *Marooned* to Academy president Mstislav Keldysh and his puzzled colleagues. As Handler later wrote of the event, the idea that "an American film should portray a Soviet cosmonaut as the hero who saves an American's life came to them as a visible and distinct shock." Shortly after the conversation, Soviet opposition to joint operations abated.[66]

Ultimately, political concerns and technical incompatibilities between American and Soviet equipment dampened enthusiasm for an intensive plan of joint flights, and the more ambitious plans survived only as a single Apollo-Soyuz experimental mission flown in 1975. Despite the optimistic start, language and culture clashes were constant in the lead-up to the ASTP flight in 1975, boding poorly for future international cooperation in space. NASA had borne the brunt of the engineering responsibility for the mission; it was the more maneuverable Apollo spacecraft that had altered its orbit to meet the Soyuz, and the Apollo's Saturn IB launch vehicle that had brought along the docking adapter that enabled crew transfers between the two craft. Soviet hardware, the Apollo advance team discovered, was at least five years behind that of Apollo, itself a seven-year-old technology.

A Place for Science: Selling the Space Shuttle

By 1977, *Star Trek* fans were no longer certain that NASA's future in space would take the form of a peaceful interplanetary federation or an internationalized interstellar space cruiser. Apollo technology was experimental, limited in capability, and dominated by white male American military aviators. What, instead, should spaceflight look like in the 1980s? *Star Trek*'s fans had provided the answer: in the future universe described in *Star Trek*: the navy, not the air force, had taken charge of human spaceflight and instead of solitary fighter pilots penetrating the void, Star Fleet's starships sent thousands of helmsmen, scientists, engineers, and even ordinary people on grand adventures to exotic worlds filled with diverse amusements and amorous alien women. Skylab and Apollo-Soyuz were, at best, imperfect efforts to redirect NASA's efforts. Crew sizes remained small, and opportunities for nonpilots in the space program remained limited. NASA would need to find a new paradigm in exploration.

Had Congress and the Nixon administration accommodated its budget requests, NASA would have constructed, in the 1970s and 1980s, a series of space vehicles, space stations, and interplanetary craft as part of a broad program of exploration. Of these proposals, only the space shuttle survived the Nixon administration's cost-cutting. Though itself only capable of flight into low-Earth orbit, the space plane's large size and apparent versatility suggested that it would not only keep the United States flying people in space for the foreseeable future, but perhaps also satisfy some of the diverse constituencies—from environmentalists to science fiction fans—still lobbying for piloted space flight.[67] In announcing the space shuttle program on January 5, 1972, President Nixon pandered to every community Project Apollo had neglected, claiming that the same vehicle that would reduce the cost of spaceflight could satisfy the nation's diverse space enthusiasts while simultaneously satisfying environmentalists, and peace activists, and civil rights advocates.

Before describing its features in January 1972, Nixon mollified human spaceflight supporters by promising that the shuttle would ensure a viable space program for decades to come, and, with a nod to space station supporters, a "real working presence in space." The true conquest of space, Nixon cautioned would happen "in the 1980's and '90's," but the space shuttle of the 1970s would be a step in the right direction.[68] Those favoring a broadening of access to space among women, minorities, and scientists, Nixon suggested, would benefit as well from the new shuttle, as "changes in modes of flight and re-entry will make the ride safer, and less demanding for the passengers, so that men and women with work to do in space can 'commute' aloft, without having to spend years in training for the skills and rigors of old-style space flight." Even the Department of Defense (DoD), Nixon, announced, would benefit from the cargo-carrying capability of the new vehicle, but hippies would be welcome, too; Nixon ended his speech with an appeal to internationalism, and "the imperatives of universal brotherhood and global ecology," promising that "This new program will give more people more access to the liberating perspectives of space, even as it extends our ability to cope with physical challenges of Earth and broadens our opportunities for international cooperation in low-cost, multi-purpose space missions."

To the sizable antispace contingent in Congress, Nixon promised an earnest effort to "minimize technical and economic risks" with a "cautious evolutionary approach," and promised that, successful or not, the program would ensure "robust activity in the aerospace industry" and the "continued pre-eminence of America and American industry in the aerospace field." As a salve to advocates of unpiloted probes, Nixon promised that the with the shuttle, "limiting boundaries between our manned and unmanned space programs will disappear" and that it would replace a host of expendable launch vehicle and robotic systems, offering a capability NASA had never before possessed, in orbit "[r]epair or servicing of satellites...as will delivery of valuable payloads from orbit back to Earth." By ferrying satellites as well as people into orbit, Nixon promised budgetary critics, the shuttle would "take the astronomical costs out of astronautics." This promise, that the shuttle would do more with less, would prove the most vexing to keep.

Unaccustomed, during Apollo, to making technical sacrifices to secure political support, NASA managers found themselves forced to negotiate for funds, manipulating cost estimates to curry favor with skeptical politicians. Unable to raise the $10–13 billon required for shuttle development, NASA management offered a $5 billion version supported by cost estimates that were demonstrably false, even by NASA's own calculations. Eventually burdened with a vehicle it could not afford, NASA channeled an increasing share of its resources into routine shuttle operations, slowing innovation and diluting its character as a research and development organization. Risks that NASA once accepted as normal within the context of a research program became unthinkable in a program promising routine access to space.[69]

To offset some of its costs, NASA announced plans to launch commercial satellites aboard the shuttle, turning the vehicle into a "space truck" that would deliver parcels for hire. The "space truck" language was too ungainly to attract public favor; the term, with its boring connotations, was misinterpreted by the press and proved a public relations headache.[70] Scientists, NASA struggled to explain, would be, instead, the primary beneficiaries of the new program.[71] On some missions, a specialized laboratory compartment—Spacelab—would fly,[72] nestled in the shuttle's cavernous cargo bay, to provide laboratory space and equipment for crews of international specialists conducting scientific research. Skylab, too, had been built to serve as a center for research, but the space shuttle was designed and constructed specifically to carry larger numbers of scientists than any earlier vehicle could accommodate. Skylab's scientific crew was limited by the operational constraints of the Apollo ferry vehicles and NASA's aversion to science pilots. The shuttle, though, would be purpose-built to accommodate a predominantly nonpilot crew of scientists freed from flying responsibilities and able to devote themselves fully to scientific pursuits.

To an organization that had defined itself by its exclusivity, the conception of the astronaut as a trained pilot, a vehicle intended principally for passengers—and frequently likened to an airliner—appeared to entail radical restructuring of the Astronaut Corps.[73] On one hand, the proposed spacecraft appeared to offer crew positions for untrained personnel (who could easily be women or minorities excluded from test pilot schools), but the idea of the shuttle as a passenger vehicle open to all was anathema to an agency founded upon the technical competence of its personnel. In the immediate future, though, NASA had no plans to expand its astronaut ranks, so it found itself publicly marketing a different craft than the one it planned to build. As if to seal the value of popular fantasy in its creation, when the first space shuttle Orbiter rolled out of the factory in 1977, a letter-writing campaign forced NASA to change its the name from *Constitution* to *Enterprise*, the vessel that propelled the *Star Trek* cast "where no man has gone before" (Figure 5.2)[74]

Criticism of the shuttle from outside NASA was immediate and unrelenting. Controversial former scientist-astronaut Brian O'Leary was among the most vocal opponents, authoring several opinion pieces in the *New York*

Figure 5.2 NASA administrator Dr. James Fletcher (left) and Star Trek cast members DeForest Kelley, George Takei, James Doohan, Nichelle Nichols, Leonard Nimoy, Star Trek creator Gene Rodenberry, and Star Trek cast member Walter Koenig stands with space shuttle Orbiter Enterprise in Palmdale, California in 1976 (NASA photo)

Times that attacked the decision to develop the vehicle, questioned its role, and suggested that cost estimates were fanciful. Recounting the history of the program, O'Leary noted that NASA had developed the shuttle to service a permanent orbital outpost to be used a way station for interplanetary flight; two goals that NASA had recently disavowed. Instead, NASA positioned the shuttle as "an end to itself"—an economical vehicle to launch satellites into space and subsequently recover and repair them.[75] O'Leary and others doubted the economy of such a plan: existing unpiloted launch vehicles were amply capable of managing the existing launch traffic, and space technology changed too quickly to justify the retrieval of NASA's aging satellites.[76] Shorn of a proper mission, O'Leary feared, "the shuttle will be used for manned stunts of little intrinsic value."[77] Born at the height of the Cold War, human spaceflight had appeared to have outlived its usefulness, becoming less a "crash program" than a permanent technoscientific mobilization maintained for reasons few could adequately articulate.

To O'Leary, NASA's public pronouncements about the space shuttle's progressive goals served as smokescreens for a vehicle whose design characteristics were guided by DoD requirements (including a payload bay large enough to accommodate military satellites), yet for which DoD would pay nothing.[78] By 1972, NASA had already become keenly sensitive to the DoD's likely participation in future shuttle projects, accepting military pilots from the air force's defunct space station project into the astronaut corps in 1969 largely to develop good working relationships with air force and navy leaders with whom it would need to cooperate to secure continued funding with the shuttle. NASA had always enjoyed relative freedom from the overt militarization of its efforts, and, supporting the shuttle, the DoD wanted more than merely temporary accommodations: it would eventually ask for a direct hand in selecting astronauts, a role it had never played, and one that seemed to contradict the growing pressures on NASA to democratize the astronaut pool.[79] As O'Leary feared, the air force, seeking an opportunity to provide crews on military missions, eventually recruited a class of military specialists to fill out the crews of DoD shuttle flights: the active-duty military Manned Spaceflight Engineer (MSE).[80]

In justifying space shuttle to the public, NASA administrator James Fletcher, in 1972, had argued that it represented an effective platform for international science programs; the laboratory module it would carry, eventually dubbed Spacelab, would be built in Europe and eventually staffed by specialists who would fly on particular missions to support their experiments.[81] Despite NASA's claims that the scientists would be welcome aboard the new craft, the role of scientist-astronauts in the shuttle, even with the promise of more laboratory space and more frequent flights, was unclear. Having arrived with few expectations, and having been conditioned by Deke Slayton not to expect a spaceflight, many scientist-astronauts immersed themselves in support work, often enjoying themselves.[82] Yet the space shuttle, astronaut Philip Chapman feared, would not only cement labor differences between pilot- and scientist-astronauts, but also flatter the pilots' preoccupations and emphasize their skills over those of the more broadly trained scientists.[83] At least in the short term, it was unclear whether NASA would require the services of even its existing scientist-astronauts to crew early shuttle test flights, which would require only the services of a handful of pilot-astronauts.

Given the large number of astronauts already in the queue, NASA, by 1973, already had sufficient personnel to crew early shuttle flights and would seek no new astronauts until 1978.[84] As the least essential personnel in NASA's estimation, scientist-astronauts would be the first to be fired once Apollo ended. Each astronaut, one 1972 memo to George Low suggested, "should be released by NASA when his services are no longer required," a plan Low noted was "reasonable," and which he suggested be discussed with Robert Gilruth and Chris Kraft.[85] Though NASA never implemented the mass firings, the role that scientist-astronauts would play on the shuttle was less well-defined than that of the pilot-astronauts who would constitute the shuttle's flight crew, and their future in NASA was, therefore, the least certain.

On September 20, 1973, Low forwarded a letter to Frederick Seitz, chairman of the NASA's Space Program Advisory Council (SPAC), requesting that the SPAC establish an ad hoc subcommittee to "undertake a review of the effectiveness and value of NASA's scientist-astronaut program in the past, and to make recommendations on the use of scientists astronauts in the space shuttle program."[86] The subcommittee, chaired by Homer Newell, presented its report the following March to Seitz, who forwarded it to Low, attaching a brief letter summarizing the report's conclusions, especially that NASA continue and expand the scientist-astronaut corps in consonance with space shuttle science and application needs."[87]

An articulation of the expectations of NASA's scientists for the shuttle program, the report was modest in its recommendations, and in some respects confused rather than clarified the issues. Despite the fact that more of NASA's scientist-astronauts had retired in disgust than flown in space, they were now an "essential part of the total astronaut team" and would remain so in the future. The only example of their necessity, though, was provided in "communications between the astronauts in space and scientists on the ground." For the shuttle project, though, the subcommittee anticipated a substantial growth in the astronaut pool, and articulated three categories of new scientist crew members with overlapping duties. NASA's existing scientist-astronauts would continue to fill crew seats, but only to support "a wide variety of scientific experiments" and would serve no piloting role, despite their training.[88] Clarifying the staffing requirements for shuttle flights, Chris Kraft, in a 1974 letter to John F. Yardley, NASA associate administrator for Manned Space Flight, found himself unable to endorse the scientists' continued piloting role. Defining the space shuttle as an experimental aircraft with "demanding stability and control characteristics," Kraft insisted that only aviators with "considerable flight test experience" would be considered as shuttle pilots, at least during the earliest flights.[89]

Instead of providing a vehicle designed for scientist-astronauts, the space shuttle would prove inhospitable to them. Indeed, NASA's former scientist-astronauts would have to compete for crew seats even with a variety of other personnel,[90] especially a new group of "mission specialist" astronauts that NASA would recruit for shuttle missions but provide with no pilot training.[91] Without any need for the scientist-astronauts' special skills, Apollo's remaining "space pilots" could be easily folded into the Mission Specialist category. At a January 1975 meeting, Fletcher, Low, Yardley, and others determined that the term "scientist-astronaut" would be discontinued, and the men transferred to the new department.[92] In September 1976, Kraft forwarded a letter to NASA assistant administrator John Naugle confirming that NASA's remaining scientist-astronauts had been assigned to the new Office of Mission Specialists headed by Kerwin, effectively demoting them and ending their piloting careers. Kraft noted that "we will not train the specialists to be pilots or teach them how to fly the shuttle." Instead, Kraft promised more support for the astronauts' scientific activities so that, while remaining "qualified as shuttle crew members," Mission Specialists would be permitted to "keep active with their peers and stay involved in research. . . ." "In short," Kraft concluded, "I want to give the Mission Specialists visibility in the Shuttle Program; I want them trained to do their job on a particular flight; and I want them competent and content professionally." [93] If pilot-astronauts coveted the rare stick time afforded them in the shuttle, though, NASA's scientist-astronauts chaffed against program management that, despite public pronouncements of support, accorded them less respect and addressed scientific interests in a haphazard fashion. The space shuttle would at last provide a large number of crew positions for scientists, but at the cost of a sharp division of labor between the pilot- and scientist-astronauts and a decline in status of the latter.[94]

In the same way that NASA promoted the shuttle as a vehicle for scientists while simultaneously demoting them, NASA continued to describe the shuttle as a vehicle for passenger travel, despite having no plans to facilitate that goal. Joining NASA's mission specialists on shuttle flights, would, in theory, be a variety of less-well-trained, noncareer astronaut "payload specialists," private individuals representing the payload sponsor: an agency, institution, or foreign government that had paid to fly its cargo on the shuttle. As described in the SPAC report, the payload specialists would exert managerial control over aspects of the mission and over NASA's career mission specialists; mission specialists would have no executive responsibilities and would assist with mission activities only at the "discretion" of the "payload sponsor."[95] Not only would NASA fly noncareer astronauts for the first time in the shuttle, the SPAC envisioned circumstances in which these individuals might command better-trained career NASA astronauts. Demoted and without any immediate use for their training, several scientist-astronauts sought opportunities elsewhere, both within and outside NASA. Ed Gibson, fresh from Skylab, was discouraged by the long wait he would face for a second flight and resigned in November 1974, accepting work in private industry. Joe Allen accepted an administrative post as NASA assistant administrator for Legislative Affairs in Washington, DC. The eight scientist-astronauts still on active duty accepted ground duties developing scientific hardware for later shuttle operations, hoping for flights once the shuttle had completed its "experimental" flights and became fully "operational."[96]

By the fall of 1976, though, NASA management confronted criticism "throughout the agency" about the amorphous and seemingly overlapping duties of mission and payload specialists.[97] Most troubling were allegations out of JSC that NASA was willing to compromise safety and squander its human capital by selling shuttle seats on the open market. Indeed, to cultivate paying customers in the corporate sector, NASA would offer flying seats to nations and companies that "purchased a half or more of a shuttle mission," a "marketing gimmick" that to Manned Orbiting Laboratory (MOL) transfer Hank Hartsfield allowed unqualified but well-connected private citizens to take seats from career astronauts.[98] "[A] lot of us didn't like that at all," Hartsfield described, referring to his fellow astronauts, "some of us worried about these short-termers coming along and joyriding… ."[99] Not surprisingly, the most vocal criticisms emerged from astronauts and sympathetic managers like Kraft, who preferred that NASA limit flying positions to existing astronauts instead of recruiting new personnel.[100] More supportive words came from NASA's scientific community, especially NASA Ames Research Center director Hans Mark, who sought the flying opportunities for the maximum number of scientists.

Key NASA managers debated the issue in an October 1976 meeting in which Fletcher took charge of the discussion, which quickly turned away from the role of the Payload Specialist and toward that of the mission specialist, first asking whether NASA even needed them in the first place, and then whether Payload Specialists duplicated their responsibilities. Fletcher recognized that both Kraft's and Mark's concerns could be accommodated provided NASA never flew a payload specialist twice, and sought to fill crew positions with career astronauts (mission specialists) whenever possible.[101] This compromise eventually provided the basis for NASA's shuttle crew assignment policy for the early 1980s, though debate continued as NASA contemplated the imminent selection of shuttle mission specialists and the European Space Agency's recruitment of the first foreign payload specialists.

The incipient menace posed by the payload specialists continued to concern Kraft throughout 1978. In February, Naugle wrote to Fletcher of Kraft's fear that "Payload specialists may be competing with mission specialists and could ultimately become 'professional' payload specialists without having been subjected to stringent selection requirements and extensive that the mission specialists are obliged to endure." Naugle dismissed Kraft's concerns as "parochial" and insisted that mission specialists would be considered first for any mission for which they qualified, before payload specialists.[102] This accommodation to appease Kraft, though, displeased Hans Mark and many within the scientific community, who feared the policy would further restrict scientific participation in shuttle flights.[103]

Commenting on the matter in May 1978, Kraft again wrote of a "general Agency attitude that a P[ayload]S[pecialist] is required for any flight that has experiments aboard." "This concept," Kraft declared, "is not consistent with my understanding of the roles and responsibilities of the various crew members." Kraft reiterated his preference that payload specialists only be assigned to flight "when he or she has unique knowledge or capability," suggesting that NASA wasted resources and courted disaster by trying to open crew positions to amateurs. Most experiments even in the dedicated Spacelab module, Kraft wrote in an appendix to his letter, "are of the on/off variety," and thus required no particular expertise beyond that possessed by NASA's mission specialists. Instead of leaving this role to NASA's career astronauts, though, NASA had instead recruited payload specialists as a marketing tool and provided them with sufficient general training to compete with mission specialists for crew assignments. Kraft further took offense to NASA management documents that appeared to assign to payload specialists operational control of their experiments and supervisor responsibility over mission specialists, a role Kraft found intolerable, considering the payload specialists' lack of experience and the likelihood that they would be felled by space sickness during their flight. Even worse, given the flight, it was likely that a "female Payload Specialist will fly before one of the highly publicized female Mission Specialist astronauts whom we chose through a rigorous selection process."[104] Payload specialists might thus steal a critical public relations milestone from NASA.

In his reply, Kraft concluded that by actively soliciting payload specialists, not only was NASA making promises it could not keep, but the incorporation of so many untrained crewmembers would also undermine safety. NASA, Kraft concluded, was "becoming too enamored with the public relations aspects of such selections without properly considering what we are doing practically and operationally." He continued:

> While I agree that JSC may not have complete appreciation for what the Agency must do to market the shuttle, we do have an appreciation for how to make a shuttle mission successful and safe. Operations and program marketing are not necessarily compatible factors and must be carefully integrated to insure mission success. This, I believe is not being done and, as such, promises are being made now that cannot be kept when we fly.[105]

Eventually, NASA lobbied for its mission specialists to assume operational control of all science payloads, but through the late 1970s the duties of the mission specialists remained unclear.[106]

Anticipating the arrival of skittish nonprofessional crew members on American space vehicles, NASA administrator Robert Frosch, in 1980, took the step of authorizing a regulation that would extend to future space shuttle commanders the same authority to arrest troublemakers as that enjoyed by airline pilots. Citing the risk of "assault in space, perhaps due to incompatibility," NASA's general counsel had cited the threat of lawsuits that might result if a shuttle commander were forced to restrain a violent Payload Specialist, crew members who were expected to be relatively untrained and possibly foreign. Veteran astronauts asked to comment on the story responded with thinly veiled disgust: an article in the *Washington Post* quoted Apollo veteran Gene Cernan, suggesting that he "never felt the need for a written regulation or the need for brute force to get things done," while Slayton speculated about locking unruly astronauts in the shuttle's bathroom, a nook that was not provided with a door. The European Space Agency, concerned about exposing its astronauts to American law, objected to the rule.[107]

A New Kind of Astronaut

Despite open questions about how the shuttle's diverse crew members would work together, the prospect that NASA might soon fly women and minority payload specialists galvanized the agency to accelerate efforts, stalled since 1972, to recruit career astronauts other than white males. Beginning in 1959, several experiments and public relations projects had investigated women's physiological suitability to space flight. Air force brigadier general Don Flickinger had researched the potential for women to fly in Project

Mercury, but, finding little endorsement from his superiors, transferred the work to the private Lovelace Clinic.[108] Despite the later examination of 19 women flyers at the Lovelace clinic in 1960 and 1961, none of the women met the strict professional qualifications required for selection as astronauts and neither NASA nor the military services ever endorsed their recruitment, selection, or training. Almost immediately, this aspect of the human spaceflight program provoked controversy. Some in the spaceflight and aviation medicine community had favored consideration of female space candidates due to their lower mass, but the silence the suggestions received from government quarters only demonstrated how far removed scientific experts were from the decision-making process on astronaut selection.[109]

The development of a rigid institutional pipeline for astronauts in the 1960s continued to limit opportunities for women and black male applicants.[110] (During the 1960s, more female monkeys flew in space than female humans, a point not lost on commentators.[111]) The Soviet Union's high profile 1963 orbiting of female cosmonaut and nonpilot Valentina Tereshkova did not appreciably impact the debate: Tereshkova was the lone women in the Soviet cosmonaut corps and never flew again. Though a propaganda victory, her flight, NASA managers argued, did not justify the time and expense of seeking out women astronauts for a similar "stunt" while NASA faced an accelerated timetable for lunar flight. Committed to lunar landing, NASA subordinated all other human spaceflight objectives to that goal, including equal-opportunity recruiting.[112]

Throughout the 1960s, NASA's work culture remained far from fully integrated by sex. Instead of colleagues, women appeared in the astronauts lives as wives or girlfriends, where they served as valuable public relations tools or, as Werher von Braun once described (supposedly quoting Robert Gilruth), "recreational equipment."[113] Even female support staff at NASA were relatively rare; "Nurse to the Astronauts" Dee O'Hara recalled that NASA seemed to want as few women around as possible.[114] It is unclear if male astronauts would have welcomed (as some of them suggested) women pilots who possessed the necessary qualifications;[115] their selection, Collins writes, would have required technical changes to hygiene equipment and would have produced embarrassment on the part of their male colleagues at the thought of sharing close quarters (a sensitivity, oddly, not attributed to the women). The 1966 Selection Board "breathed a sigh of relief that there were no women," noted Collins, "because women made problems." Astronauts, though, were never quoted making any statements hostile to the presence of nonwhite male astronauts. "The absence of blacks was a different matter," Collins continued, "our group would have welcomed them, and I don't know why none showed up."[116]

If the NASA was predominantly male in 1961, it was also predominant white. African Americans constituted 2.5 percent of NASA's ground personnel in 1965, but black, Latino, or Asian American pilots, like women, could be found nowhere in the astronaut ranks throughout the 1960s, a political issue that caused repeated consternation within, and frequent criticism of, the agency. Early voices for the inclusion of "Negroes" in the astronaut corps were louder and enjoyed wider support inside and outside the government than those for women, but with military test pilots the only applicants considered for space piloting, no African American pilots possessed the necessary credentials. This displeased advocates of integration who, while concerned principally with issues of fundamental fairness, deployed political arguments to support the selection of black astronauts. Summing up the opinion of many, journalist Edward R. Murrow had written to NASA administrator James Webb in September 1961 on the likely the impact of American space achievements on the "whole non-white world, which is most of it," were one of America's new spacemen to be a "qualified Negro."[117] Positioned as a demonstration of American scientific altruism, the space program stood to gain more ground among developing nations recently emerging from European colonialism if its new heroes were other than white. Murrow was not alone in his sympathies; Vice President Lyndon Johnson, too, was discouraged by the lack of a black astronaut in the early selections.[118]

Unlike the tepid response of political leaders provided to the issue of women astronauts, racial integration of the space corps received substantial encouragement at the executive level during the 1960s but was no more successful, mostly due to the paucity of nonwhite test pilots. During the 1960s, only two black pilots met the criteria for selection by NASA: Edward J. Dwight of the air force entered the evaluations for NASA's third group of pilot-astronauts in 1963, and, while he succeeded in the opening rounds of the selection, he was not among those chosen.[119] Dwight's experience became a significant point of controversy, not among astronauts (who largely avoided comment on the presence of women or minorities in the space service), but from the public and the air force. Dwight's career in the air force had been aided by a 1962 directive from President Kennedy to integrate the air force's Test Pilot School, and Dwight, a capable bomber pilot, had been selected for the course shortly thereafter. Though graduating in the middle of his class at Edwards, the taint of preferential treatment had followed him in the air force; while Dwight did not blame NASA for his failure to achieve a slot in the astronaut class, he regarded his treatment by the air force as discriminatory.[120] Robert H. Lawrence, an air force test pilot with a PhD in nuclear chemistry, died in a 1967 jet crash shortly after his selection as a military astronaut under the DoD's MOL project. Had Lawrence lived, he likely would have transferred to NASA upon MOL's cancellation and flown early space shuttle missions.

As long as vital national security interests motivated the human spaceflight program, NASA examiners like psychologist Robert Voas could readily deflect widespread criticism from Congress, nonprofit organizations, members of the press, and even Communist propagandists by arguing that the technical demands required race-/sex-/religion-blind astronaut selection criteria that inadvertently disadvantaged female and black applicants.[121] Astronauts were occasionally asked to defend NASA's apparent discrimination; they had so often defined their competence in objective technical terms—as John Glenn and Scott Carpenter had while testifying before Congress in July 1962—that they could appear race- and gender-blind without the incurring the slightest allegations of bigotry. (Suggestions that more qualified candidates had not been chosen on account of race or gender bias likely would have been extremely offensive to men inclined to view themselves as the nation's best.) If the astronauts possessed considerable power within the space program, though, they chose not to use it to ensure diversity in the Astronaut Corps. In all likelihood, such decisions were out of their hands, and it is unclear if the expenditure of the astronauts' considerable political capital would have sped the arrival of minority astronauts. Ultimately, only the increased numbers of women and minorities in the flying ranks of the armed forces, combined with civil rights initiatives, reduced the astronaut corps's homogeneity; as significant, though, was the eventual demise of Project Apollo, which had created the urgency that had kept the astronaut selection criteria so stringent.

By the time the Apollo 11 astronauts had landed on the Moon in 1969, a growing community of dissent had emerged, for whom America's space success belied a space agency barely integrated by race and gender, even by 1960s standards. Particularly problematic for NASA in the early 1970s was the continued gender and racial exclusivity of its astronaut ranks: the Moon Race had been one, but NASA would still fly only white male pilots. Without the Moon Race to shield it, the ethnic and gender homogeneity of NASA's astronaut corps also suggested a dissonance between the goals of Apollo (its obsession with putting "Whitey on the Moon"[122]) and the needs of a nation increasingly inclined to view persistent social discrimination as the leading national concern. Even *Star Trek's* USS *Enterprise* had enjoyed a crew integrated by gender and ethnicity, and while the ethnic uniformity of the military pilots of 1959 attracted relatively little controversy amid the urgency of the early years of the Space Race, political criticism had mounted in the late 1960s among commentators who demanded that if the United States were to fund spaceflight at all, it might as well have an astronaut corps more representative of the nation.

Science fiction, Amy Foster writes, had long anticipated the integration of space crews, with space fantasies throughout the mid twentieth-century, placing women alongside men aboard interstellar vessels, though often as scantily clad assistants or objects of lust. The most progressive of the late 1960s products, like *Star Trek*, though, contemplated women as bridge officers with duties comparable to those of their male colleagues, but social attitudes of the distant future seemed to parallel contemporary mores, with the males still surprised to find women "on the bridge."[123] Only a few short years later, though, years corresponding with the birth of the modern women's movement, cultural barriers to female participation in spaceflight appeared increasingly archaic, if not unlawful.

A month after *Star Trek* fans celebrated Gene Rodenberry's progressive—and, ironically, defunct—space vision, NASA administrator James Fletcher, addressing an Equal Employment Opportunity conference at Kennedy Space

Center, promised that the space shuttle would at last resolve one of the more vexing political issues that had confronted NASA during the previous decade, the difficulty encountered by women and nonwhites to enter the astronaut ranks of NASA's "manned" spaceflight program. NASA, effectively, had no choice; what had been a logistical issue in 1961, was, by March 1972, a legal one, with the passage of the Equal Employment Opportunity amendment to the 1964 Civil Rights Act, extending that law's gender and race-based antidiscrimination protections to employees of the federal government. That year, the armed forces opened test pilot schools to women, promising the possibility that women might join the piloting ranks of NASA's career astronauts.[124] In 1973, NASA renamed the Manned Spacecraft Center in Houston for former president Lyndon Johnson, though the term "manned spaceflight" remained in wide use for the next three decades.[125]

The lack of women with test pilot experience had, in 1961, proved fatal to efforts by aviators like Jacqueline Cochrane and Jerrie Cobb to prod NASA into recruiting women as astronauts; a lack of black test pilots had similarly hampered more aggressive efforts to recruit black astronauts. Criticisms of the racial homogeneity of the astronaut corps that had been tempered during the Moon Race exploded in the summer of 1972 during Apollo's decline, with a letter from Congressmen Charles Rangel (D-NY) to the Federal Civil Rights Commission, requesting an investigation of NASA. Three years later, Senator William Proxmire (D-Wisconsin) conducted a hearing on the controversy.[126] Having selected its astronauts in a closed procedure that flouted Civil Service rules, NASA now found itself struggling to demonstrate the legitimacy of its irregular, often exclusive hiring practices.

The renewed movement toward a diverse space corps, though, came, for NASA, at yet another "worst possible time," with Apollo ending and rosters already filled with astronauts from the 1966, 1967, and 1969 groups. During September of 1972, NASA management debated a new astronaut selection, with an eye principally toward maintaining sufficient pilot numbers to support flights of future vehicles.[127] By the end of December 1973, though, the press had already become impatient for progress on NASA's shuttle-related integration efforts. A letter to George Low from Brian King of the Associated Press inquired about recent NASA experiments involving air force nurses, asking about the qualifications of the test subjects and the project's apparent secrecy, as well as the possibility that the untrained "wives of astronauts" might soon join their husbands on shuttle missions.[128] Low quickly silenced the rumors, reiterating that the shuttle program would answer the concerns of critics regarding the inclusiveness of NASA's human spaceflight operations, but would, for the foreseeable future fly only technically qualified personnel.[129] With no missions for new astronaut recruits to fly in the immediate future, NASA was unable to capitalize on public sympathy for progress on equal-opportunity astronaut recruitment and Fletcher was forced to postpone the new selection until 1978.[130]

Throughout the 1970s, the contradictions imposed by the shuttle's proposed crew manifests had perplexed NASA staff: while NASA's leadership assured its Johnson Space Center in Texas that crew members would be highly trained pilots, NASA promised its Ames Research Center in California that some would be scientists, and suggested to the public that many might be women, ethnic minorities, or average civilians. At the same time that it recruited pilot-astronauts, NASA announced that such individuals were an endangered species. "This is a major change," the *New York Times* wrote, quoting George Abbey (the powerful director of the Flight Crew Operations Directorate at JSC), "because the pilot is moving into the background," to be replaced by mission specialists "who will do the work on upcoming flights." [131] These mission specialists, though, were not to be specialists at all, but generalist technicians and engineers who would assist on a space-available basis. They would compete for seats, furthermore, with relatively untrained, non-NASA payload specialists, sponsored by organizations that had paid for the privilege. On certain flights, ordinary passengers might fly as well. Against a backdrop of conflicting constituency, virtually no one could predict who might crew any particular mission.

For its pilot-astronauts, NASA continued to recruit test pilots from the military services. The new mission specialists, though, would be drawn from both PhD scientists and engineers who the 1975 SPAC study described as unable to qualify under existing categories, but whose experience made them capable to assist in various in-flight activities not directly connected to research, including space walks and satellite deployment. The ongoing lack of diversity in test pilot schools made it unlikely that NASA would find women and black pilots or even scientists in sufficient numbers, so rather than limiting the available pool of applicants, JSC's integrated plan of March 1975 provided for a broad lowering of application requirements in mixed selection that recruited both pilots (likely to remain all-white and all-male for the foreseeable future) and an ethnically and gender-diverse group of scientists. Once it had selected them, NASA would subject the new arrivals to a more intensive training regimen until they met required standards.[132]

Despite continued debate on the role of mission specialists, NASA continued with its efforts to recruit them. Instead of vetting candidates through the NAS, NASA would solicit all interested parties and cull the applications itself to ensure a sufficient number of nonwhites and women remained. Recruiting qualified women and minority applicants, though, was particularly difficult; the career NASA projected—tending experiments in an orbital space plane flown by military personnel—lacked the visceral thrill of Apollo, and many would-be applicants doubted NASA's commitment to equal opportunity employment. In order to increase the applicant pool, NASA eventually turning to Nichelle Nichols, the actor who played Lieutenant Urhura in *Star Trek*, to assist in their efforts. From the 8,000 applications received by the June 30, 1977 close, NASA selected 187 men and 21 women for medical and psychological screening. It announced its choices the following year: 15 pilot-astronauts and 20 mission specialists. Among the pilot-astronauts, competition was fierce; since 1963, NASA had welcomed nontest pilots in astronaut selections, but all of 15 of the pilots in the 1978 group were active duty military personnel with test pilot experience, and all were white males. According to one account, controversy had plagued the selection, with JSC seeking five more pilots, only to be countermanded by NASA Headquarters, which preferred to select five more mission specialists instead, including four women. The eventual arrival of the first women astronauts was easily the most dramatic transformation of the Astronaut Corps, not only reversing decades of gender imbalance but, eventually, challenging NASA to reconfigure equipment and operations to suit female bodies. Certain NASA managers, Amy Foster recounts, initially either chaffed at the presence of women or clung to expectations regarding dress and deportment, producing conflicts often rectified by Dr. Carolyn Huntoon, the lone woman on the NASA's 1978 Selection Board and a tireless advocate for the burgeoning women's astronaut corps.[133]

Democratizing Spaceflight

NASA had hoped to salvage human spaceflight by promising broad public participation in the space program, but, to do so, it had to break the pilot-astronauts' stranglehold on flight assignments and open crew positions to noncareer personnel. In 1978, NASA selected a new group of astronauts while questions still remained as to the respective roles of the shuttle's various crew members; to many at JSC, noncareer astronauts were mere dilettantes only required in the unlikely event that a particular piece of shuttle hardware so overwhelmed the crew that additional expertise was necessary to operate it. As NASA struggled to attract attention to the shuttle among foreign countries and the DoD, crew positions became valuable currency in negotiations. NASA, having heavily promoted the shuttle as a vehicle that would democratize space travel, was now faced with the problem of making good on its claim that career pilot- and scientist-astronauts, civilian technicians, military engineers, and foreign experts would all share seats on the same vehicles. Meanwhile, it aggressively promoted the idea that before too long, average citizens would join space crews, erasing the distinctions between NASA's astronauts and ordinary passengers who might receive little training, and could be almost anyone.

Reflecting on NASA's astronaut recruitment and training protocols of the 1960s, Chris Kraft, in a 1974 letter to John Yardley, stated the case for an astronaut selection process for the shuttle program that would ensure "motivated" and "personally dedicated" pilots and fly them quickly after selection.[134] Such a cadre never materialized. Instead, NASA loosened qualifications to broaden access, and quickly flooded its ranks with more astronauts than it could fly.[135] In fact, the relative comfort and sophistication of the shuttle, and the lack of anything for most nonpilot crew members to do during launch, reentry, and landing, eventually created an ideal opportunity for relatively untrained personnel—teachers, journalists, politicians, retired space heroes, foreign dignitaries—to fly in space, further diluting the exclusivity of the astronaut corps and a key element of its professional appeal.[136]

Slayton, head of the shuttle program in the late 1970s but losing ground to NASA management, protested the growth in the size of the astronaut corps and the large number of astronauts to be flown on each shuttle mission. He preferred a corps half the size, filled with veteran teams that would fly regularly and routinely, developing diverse competencies.[137] Ultimately, though, Slayton was unable to convince NASA management to accept his alternative, and a professional cadre of long-serving, highly experienced crews never emerged. Instead, new selections filled the ranks emptied by departing veterans.

Though intended to capitalize popular enthusiasms, NASA ultimately proved unsuccessful at satisfying diverse expectations and shaping public opinion. Space colonies, international exploration, and environmentalism appealed to public, but NASA's new "space truck" did not, nor did NASA's efforts of the early 1980s to trade space adventure for the promise of occasional citizen participation. Americans wanted share in the glamour and excitement of space travel, a goal the shuttle would never deliver, especially after destruction of the Orbiter *Challenger* and the violent death of its crew—including the first private citizen to fly in space, teacher-astronaut Christa McAuliffe—during a highly publicized launch attempt on January 28, 1986.

Conclusion

In July 2005, a periodical calling itself "the world's only reliable newspaper" broke a story that an American military aviator secretly rocketed into space in 1958 had finally returned to Earth. During his 47 years in orbit, air force colonel "Hot Diggity" Corey had not aged a single day. In a brief exclusive, the *Weekly World News* described Corey as an uncomplicated, uncommunicative man-child, utterly oblivious to the enormity of his journey through space and time. In his debriefing, Corey merely apologized for dozing off in orbit and registered his delight at the prospect of being crewed with one of the female astronauts of the National Aeronautics and Space Administration (NASA) on his next flight. The very first question Corey had asked upon returning to Earth, though, concerned the availability of popsicles. When told that they still existed in the year 2005, the simple-minded, skirt-chasing space monkey responded simply: "Hot diggity!"[1]

Informed by the writings of Tom Wolfe, this caricature of the earliest astronauts emphasizes a combination of controversial character traits supposedly manifest by some early spacemen.[2] Corey is a lecherous dullard, confident but fallible, asking for nothing more in life than a craft to fly and a woman to romance. This image is less inaccurate than incomplete; while some military test pilots fulfilled these stereotypes, most did not. More often, the characteristics that united early astronauts were not stupidity, but intelligence, often concealed behind a veneer of bravado; not indifference, but ambition bordering on obsession; not folksy good humor, but an unwillingness to disclose too much of themselves to anyone in a position to undermine their work.

Created to serve an unusual set of technical and political needs in 1958, NASA's astronauts provide an interesting case study in the creation of a new—and, ultimately, failed—technical profession. In the early years of the space race, concerns over Soviet achievements transformed American high-altitude research from an incremental program under military auspices to a civilian program of big-budget exploration, giving Congress and the public a powerful role in shaping technology policy. Test pilots, who served as respected technical experts in the previous X-plane program, became public figures in the new National Aeronautics and Space Administration, and labored over the next five years to influence the agency that employed them.

Astronauts weren't management, but they were NASA's public face, and their words carried more weight in the public mind than those of the men who had hired them.

In their efforts to influence national space policy, astronauts were, at least at first, surprisingly successful. Within only a few short years, they insinuated themselves in vehicle design programs and assumed responsibility for recruitment, selection, training, and assignment of new spacemen. Astronauts turned Project Mercury's bare capsules into piloted space vehicles and laboratories, and made numerous technical and managerial contributions to the Gemini and Apollo programs. Astronauts accomplished these goals not through rebellion but through cagey exploitation of professional relationships and their understanding of the dynamics of large organizations. These efforts culminated in the successful completion of a series of absurdly dangerous, extremely challenging lunar flights, well within the time constraints set by national political authorities.

After the success of the Moon program, though, these once-obscure aviators learned that more distant flights were not forthcoming. By 1970, as great a mismatch in supply and demand existed for astronauts as for physicists in nuclear weapons work, another casualty of the easing of Cold War tensions.[3] Unlike physicists, astronauts had a poor selection of career alternatives. For some of the men selected in 1959, 1962, and 1963, the space program simply passed them by; others found that a single trip to the Moon ruined them for just about any other kind of work.[4] Instead of an open-ended campaign of interplanetary exploration, the men began to recognize that were they to remain at NASA, they would be participants in something quite different: a large, bureaucratically managed technical infrastructure.

At first a small, elite group of military test pilots with high public visibility hoping to managing their own affairs, NASA's astronauts found themselves, by the early 1970s, an increasingly beleaguered small-interest group struggling to maintain its authority amid a decline in the number of planned flights, the increasing bureaucratization of NASA, the retirement of the most respected senior astronauts, and a large growth in the number of fungible rookies. Senior physicists writing about the throngs of students flooding American graduate programs after World War II could complain about the youngsters' conformity and quest for a fast buck.[5] Senior pilot-astronauts of the 1960s and 1970s, too, complained about young, weak, unmotivated, "inferior" colleagues.[6] These fresh arrivals included new kinds of astronauts who undermined the veterans' carefully constructed image.

NASA's scientist-astronauts could neither control their public image as their predecessors had, nor use their visibility to force NASA to accommodate their interests. Rocketry research had often occurred amidst rustic surroundings; scientist-astronauts accustomed to culturally stimulating academic communities, though, felt isolated in Houston.[7] Some were so demoralized that they left NASA soon after joining; those who stayed found few flying opportunities.[8] Scientist-astronauts selected in 1967 waited, on average, more than five times longer for their first flight than those of the

1962 group, and twice as long as the members of the 1966 and 1978 groups. By contrast, the small group of military astronauts transferred to NASA from the Manned Orbiting Laboratory project brought with them, at least, a certain status that scientist-astronauts could never obtain despite years of preparation. Although their junior status would bar them from immediate flying opportunities, the valuable skill set they brought with them—and their determination to wait as long as necessary to fly—guaranteed them a position of importance in future projects.[9] While members of both groups eventually flew, it was the more junior pilots of 1969 who eventually commanded the more senior scientist-astronauts of 1967.

Without a Space Race to motivate continued Congressional appropriations, NASA and its astronauts flailed. Unable to rely upon continued Presidential and Congressional largesse, NASA, during the early 1970s, at first hoped that progressive interests like environmentalism might replace nationalism in justifying continued space exploration. These efforts gave brief and unexpected agency to countercultural forces that, while creative and boisterous, could not provide the same measure of financial support. For astronauts, the perks of office, the protections of the press, and the enthusiasms of the public slowly disappeared, exposing spacemen to new forms of scrutiny. Deprived of operating capital, NASA relegated some of its most impressive space vehicles to public display, especially in the Smithsonian Institution's National Air and Space Museum.

Completed in 1976, the Museum quickly emerged as a new temple to Apollo, celebrating the glory of American space achievements and linking the United States to a central narrative of optimistic exploration of the heavens. As the 1970s progressed, NASA increasingly sought to harness the wellspring of popular enthusiasm for science fiction that such missions had inspired. In the adventures of fictional space travelers of the 1970s, Americans imagined adventures in the distant cosmos, challenging NASA to reconfigure its new space shuttles as space cruisers that would both undertake exciting adventures and open space travel to average people. The space shuttle, its supporters often repeated, would produce scientific innovations merely through its existence. Mostly, though, NASA tried to create "big science" by building a big ship—visitors to the shuttle Orbiter *Enterprise* at the Stephen F. Udvar-Hazy Center of the Smithsonian Institution's National Air and Space Museum were generally shocked by the shuttle's enormity—and hoping that science would follow. Instead of ushering in a new era of spaceflight, though, NASA's astronauts selected in 1978 would look very much like their earlier colleagues, spending five years or more waiting for their first flight and retiring, often unhappily, ten years later, after only their second or third.[10]

Clogged by bureaucracy, adding astronauts to its already bloated ranks, suppressing reports of technical problems with its new space shuttle, NASA, by 1980, had become a large, cash-strapped government agency straining to maintain the national technical infrastructure amid an uncertain economic landscape. Ultimately the American human spaceflight program of the 1980s was not an organization in which most of America's veteran pilot-astronauts chose to remain. Assembled for the accelerating lunar landing initiative, this collection of talent could not sustain itself during a period of routine or diminishing flying opportunities; space veterans demanded regular promotions and more rewarding missions. When astronauts did not get them, they quit.

The "average" astronaut during America's first three decades in space was a male military aviator selected at the age of 33. Though he may have joined the space program in the hope of flying to the Moon or Mars, he would not visit either; in fact, he would never travel more than a few hundred miles above the Earth's surface. After a little over seven years of training he would have made his first flight into Earth orbit. Four years later, he would have returned to space in command of a similar mission. On those two flights, he would have logged a total of 21 days in space and would have made a single space walk lasting half a day. Upon returning from his second flight, the astronaut would not have lingered long at NASA; within a year and half after returning to Earth, he would have retired, at the age of 46.

The statistics above suggest a far more mundane conception of the astronaut than that popularized in the media. The first astronauts who joined NASA experienced celebrity and shortened careers; those selected shortly thereafter, multiple flights, lunar voyages, and professional dislocation; and those who joined the shuttle program, long waits for flights and relative anonymity. In 2006, scientist-astronaut and Mission Specialist Story Musgrave spoke of the constant need to perform at the highest possible level, mindful that "millions" could have performed his job and thus of the need to always justify his place in the flight roster. 1978 Mission Specialist Mike Mullane constantly sought to fit in with pilot colleagues, joking to conceal his lack of knowledge of the sports trivia that the pilots endlessly debated.[11] Reflecting on his years of experience training astronaut crews, noted NASA engineer Homer Hickam wrote, in one widely circulated op-ed piece, of astronauts who, forced to compete with each other for precious flight opportunities, eventually became "bureaucratic combatants with warped personalities and shaken confidence." Having arrived as "enthusiastic young astronauts" they soon transformed into "powerless, stressed-out peons within their own organization," unable to affect their professional destiny except through grueling, merciless contests to gain the attention and respect of NASA management. Indeed, some astronauts, Hickam wrote, adopted arrogant personalities and mistreated even senior ground engineers and scientists; others, he postulated, merely retreated into the background and, eventually, snapped from the pressure.[12]

The narrative of overexpansion, collapse, and rebirth in American astronautics in the 1970s, the delays endured by the astronauts of 1967 and 1969, and their disparate treatment, may be familiar to military or labor historians who study the effects of labor specialization, workplace relations, mobilization and demobilization, and boom/bust cycles. If not for the "leftover" astronauts of 1966, 1967, and 1969—an unintended byproduct of Apollo's cancellation—NASA would have found itself, in 1981, with roughly the same number of qualified astronauts it had in 1961. The astronauts who stayed through the 1970s, though, displayed a patience and resolve uncharacteristic of test pilots in a "crash program."

Ironically, just as NASA struggled with the mass departure of veteran astronauts, public interest again turned to the earliest generation of spacemen. For many Americans, the fantasy of space travel was not one of tourism but of heroism, with a new generation of media products rekindling popular enthusiasm for the earliest pilot-astronauts, beginning with Tom Wolfe's *The Right Stuff*, in 1979. In the years that followed, popular culture increasingly downplayed the identity crises of veteran astronauts and emphasized the qualities that had made them heroes in the first place, especially their seeming indifference to danger. The 1979 television iteration of the classic Buck Rogers character from American film serials of the 1930s replaced the futuristic space voyager with a space shuttle pilot-astronaut, accidentally frozen and thawed out five hundred years in the future. Once restored to health, Gil Gerard's character in *Buck Rogers in the 25th Century* (1979–1981) seamlessly transitions into Earth's future space force and instructs the unskilled pilots in twentieth-century air combat tactics. In the 1982 film, *Firefox*, an advanced Soviet fighter plane is stolen by a veteran American pilot played by Clint Eastwood. Once dispatched to Russia, Eastwood's character proves so skilled that one Soviet Air Force general assumes the thief must be a "NASA astronaut."[13] Similarly emerging from the prison of his own past in a space-themed film was 1970's film tough guy Jack Nicholson, playing a dissolute retired astronaut in 1983's *Terms of Endearment*. Though introduced as an unappealing playboy surrounded by sports cars and memorabilia—Garrett Breedlove (Nicholson) eventually wins the girl (Shirley MacLaine) and reveals a bottomless reservoir of emotional strength.[14] By 1983, America's astronauts of the 1960s had receded into popular memory so completely that the Philip Kaufman's film adaptation of *The Right Stuff* could introduce a stylized version of the Original Seven to a new American audience inclined to romanticize their personalities and exploits.

Compared to these men, America's newest space travelers were a bore, or, at worse, a stunt, for which the public had little enthusiasm. In 1994, an episode of long-running animated series *The Simpsons* blasted incompetent nuclear plant technician Homer Simpson into space under NASA's fictional program to launch a "blue-collar slob" into orbit. The mythical space shuttle that will carry him is named "*Corvair*," after the automobile that Ralph Nader famously described as "unsafe at any speed."[15] While Homer's training is grueling, his duties aboard *Corvair* are not. He is accompanied by "veterans": real-life astronaut "Buzz" Aldrin (voiced by the man himself) and a fictional pilot, "Race Banyon," from an earlier animated adventure series. The placement of the accident-prone Homer into the vulnerable space vehicle seems fraught with danger, and sooner after arriving in orbit, Homer fills the cabin with potato chips and breaks the shuttle's experimental ant colony with his bulbous, beer-soaked cranium.[16]

Why had NASA sent a lummox like Homer into orbit? Because the public had not supported the previous decades' shuttle flights. At the beginning of "Deep Space Homer," NASA managers struggle to understand the American people's lack of enthusiasm for the voyage of a previous crew comprised of "a mathematician, a different kind of mathematician, and a statistician."[17] Popular culture representations like "Deep Space Homer" emphasized popular fatigue, both with safe, dull voyages undertaken by college professors, and a space program that sought favorable press by providing once-in-a-lifetime trips to single members of select—and often privileged—professional groups. In *The Simpsons*, Homer's flight is at first a PR coup, but his achievements are soon overshadowed by those of an inanimate carbon rod that saves the crew, a joke at the expense of Homer, and of NASA's faulty hero-making effort.

Veteran spacemen emerge from "Deep Space Homer" relatively unscathed; increasingly, through the 1990s, seasoned astronauts were likely to be portrayed in the media as heroes, even as writers poked fun at their spare tires, weak eyes, and graying hair. In film, nearly everything mild about the astronauts is sanded clean, revealing the characters to be aggressive figures acting as much in opposition to NASA as in concert with it. In *Deep Impact* (1998) and *Space Cowboys* (2000), a gaggle of over-50 Hollywood leading men—Robert Duvall, James Garner, Tommy Lee Jones, Donald Sutherland, and Clint Eastwood—play test pilots returning to the cockpit to save humanity from impending disaster.[18] NASA itself inspired these tales: the handful of remaining Apollo veterans had constituted the bulk of the shuttle's commanders and pilots through the end of 1985, and by the mid-1980s, the advancing age of 1960s-vintage astronauts was already becoming something of a joke within the program. The maiden flight of the shuttle *Challenger*, 1983's STS-6 mission, brought together two 1969 pilot-astronauts and a 1967 scientist-astronauts, under the command of a 1966 astronaut who had logged a single Skylab flight in 1973, in a crew nicknamed the "Geritol bunch."[19] An unofficial mission patch distributed to crewmembers of the 1983 STS-8 mission depicted a sleepy-eyed, bespectacled Richard Truly in one darkened window of the shuttle Orbiter *Challenger*, and in the other the four pairs of frightened eyes, representing the rookies with whom Truly, on his second space flight in 14 years, would be flying into orbit. The image is a striking one, and one that the press noticed at the time; in Mercury, Gemini, and Apollo, an astronaut wearing glasses was unheard-of.[20] In 1998, 77-year old Senator John Glenn, one of America's first astronauts, even returned to space aboard the shuttle. On the flight, Glenn served principally as a biomedical experiment; Hollywood's versions seemingly exaggerate Glenn's accomplishments, attempting to prove that old, bold pilots still do exist.

While the lead characters in these films dodge age-related jokes, most of the laughs are on the veterans' fresh-faced crewmates, who prove unable to endure the stresses of spaceflight. In *Space Cowboys* in particular, younger, better-educated astronauts are ridiculed, and the excesses of first-generation spacemen—womanizing, drinking, anti-intellectualism—are celebrated.

Many of the old codgers are widowers, and their romantic exploits are played up as evidence of the veterans' authenticity and continued vitality. Whereas Eastwood, in 1992's *Unforgiven*, attempted to demythologize the brutal gunslingers he played in a string of 1960s Westerns, *Space Cowboys* is a homage to behavior NASA once tried desperately to conceal, now interpreted as gusto.[21]

If the return of "Hot Diggity" Corey nearly half a century after blast-off reminds Americans how far the Astronaut Office has matured since 1959, it also reminds them of what, perhaps, had been lost. The popular media satirized the image of the first pilot-astronauts almost immediately upon their selection, but however comical test pilots appeared, many of the nation's commentators preferred to see taciturn, courageous test pilots in space capsules over virtually anybody else. The astronaut corps after 1978 would be more diverse in sex, ethnicity, age, and experience—in short, more like American society in general—than that of the 1960s.[22] Instead of a steely pilot with a demure and faithful wife, the first archetypal astronaut of the twenty-first century was space shuttle commander, air force colonel, wife, and mother, Eileen Collins, who was celebrated in the press for her alleged maternal temperament.[23] It was 50 years after Alan Shepard's first flight that the United States enjoyed the largest and arguably most capable astronaut corps in NASA's history, and yet the space vehicle, while as dangerous as ever, often appeared as a domesticated environment seemingly devoid of adventure or high return.

Notes

Introduction

1. For example, Roger D. Launius, "Heroes in a Vacuum: The Apollo Astronaut as Cultural Icon" (paper presented at the 43rd AIAA Aerospace Sciences Meeting and Exhibit, Reno, Nevada, January 10–13, 2005).
2. For example, Jerry Bledsoe, "Down from Glory," *Esquire*, January 1973, 83–86; Brian O'Leary, *The Making of an Ex-Astronaut* (Boston: Houghton Mifflin, 1970); Edwin E. Aldrin and Wayne Warga, *Return to Earth* (New York: Random House, 1973); Michael Collins, *Carrying the Fire: An Astronaut's Journeys* (New York: Farrar, 1974); Walter Cunningham and Mickey Herskowitz, *The All-American Boys* (New York: Macmillan, 1977).
3. Tom Wolfe, *The Right Stuff* (New York: Farrar, Straus, and Giroux, 1979).
4. Walter A. McDougall, *The Heavens and the Earth: A Political History of the Space Age* (Baltimore: Johns Hopkins University Press, 1997).
5. While labor historians employ the word "autonomy" to describe workers' degree of control over their labors, the word, as used by NASA in internal memoranda, generally referred to the ability of a crew to operate in orbit independently of ground controllers. Even these activities, though, would normally follow a well-planned sequence of prescribed steps. See, for example, "Crew Autonomy," November 21, 1978, Box 1, Thomas K. Mattingly Files, Center Series, Johnson Space Center History Collection, University of Houston–Clear Lake.
6. Mindell, "Human and Machine in the History of Spaceflight," in *Critical Issues in the History of Spaceflight*, ed. Steven J. Dick and Roger D. Launius (Washington, DC: NASA, 2006).
7. Harry Braverman, *Labor and Monopoly Capital: The Degradation of Work in the Twentieth Century* (New York: Monthly Review Press, 1975).
8. Richard Edwards, *Contested Terrain: The Transformation of the Workplace in the Twentieth Century* (New York: Basic Books, 1979).
9. Walter Licht, *Industrializing America: The Nineteenth Century* (Baltimore: Johns Hopkins University Press, 1995).
10. Robert Zussman, *Mechanics of the Middle Class: Work and Politics among American Engineers* (Berkeley: University of California Press, 1985).
11. Monte A. Calvert, *The Mechanical Engineer in America, 1830–1910: Professional Cultures in Conflict* (Baltimore: Johns Hopkins University Press, 1967).
12. Magali Sarfatti Larson, *The Rise of Professionalism: A Sociological Analysis* (Berkeley: University of California Press, 1977), xvii.

13. See, for example, Andrew Delano Abbott, *The System of Professions: An Essay on the Division of Expert Labor* (Chicago: University of Chicago Press, 1988), 59.
14. Michael L. Smith, *Pacific Visions: California Scientists and the Environment, 1850–1915* (New Haven: Yale University Press, 1987).
15. Chandra Mukerji, *A Fragile Power: Scientists and the State* (Princeton: Princeton University Press, 1989).
16. Asif A. Siddiqi, *The Rockets' Red Glare: Spaceflight and the Russian Imagination, 1857–1957* (Cambridge, UK: Cambridge University Press, 2010).
17. Sheila Jasanoff, "Image and Imagination," in *Changing the Atmosphere: Expert Knowledge and Environmental Governance*, ed. Clark A. Miller and Paul N. Edwards (Cambridge, Massachusetts: MIT Press, 2001).
18. Compare, Helen M. Rozwadowski, "Small World: Forging a Scientific Maritime Culture for Oceanography," *Isis* 87 (1996) (scientists as undesirable passengers aboard maritime vessels.)
19. Robert W. Farquhar, "Fifty Years on the Space Frontier: Halo Orbits, Comets, Asteroids, and More" (unpublished manuscript, 2009), ii.
20. Larson, 149.

1 "Project Astronaut"

1. Loyd S. Swenson, James M. Grimwood, and Charles C. Alexander, *This New Ocean, a History of Project Mercury* (Washington, DC: NASA, 1966), 132.
2. See, for example, Craig Ryan, *The Pre-Astronauts: Manned Ballooning on the Threshold of Space* (Annapolis: Naval Institute Press, 1995).
3. See David A. Mindell, "Human and Machine in the History of Spaceflight" in *Critical Issues in the History of Spaceflight*, ed. Steven J. Dick and Roger D. Launius (Washington, DC: NASA, 2006), 149.
4. Michael Collins, *Carrying the Fire: An Astronaut's Journeys* (New York: Farrar, 1974), 14–15.
5. George T. Haughty, "Human Performance in Space," in *Man in Space: The United States Air Force Program for Developing the Spacecraft Crew*, ed. Kenneth Franklin Gantz (New York: Duell, 1959), 85.
6. Mindell, "Human and Machine in the History of Spaceflight," 148–149; David A. Mindell, Digital Apollo: Human and Machine in Spaceflight (Cambridge, Massachusetts: MIT Press, 2008), 29–32.
7. "Testy Test Pilots Society (for Lack of a Better Name at the This Point): Minutes of the First Organized Meeting," in *History of the First 20 Years*, ed. Society of Experimental Test Pilots (Covina: Taylor Pub. Co., 1978), 10; Society of Experimental Test Pilots, *History of the First 20 Years*, 11; Mindell, *Digital Apollo*, 31.
8. Robert Gilruth, "Transcript #5 (David DeVorkin, John Mauer, Interviewers)," Oral History Transcripts (Washington, DC: Smithsonian Institution, National Air and Space Museum, February 27, 1987), 4.
9. See, generally, Roy Franklin Houchin, II, "The Rise and Fall of Dyna-Soar: A History of Air Force Hypersonic R&D, 1944–1963" (PhD dissertation, Auburn University).
10. Robert Gilruth, "Transcript #4 (Martin Collins, David DeVorkin, Interviewers)," Oral History Transcripts (Washington, DC: Smithsonian Institution, National Air and Space Museum, October 2, 1986), 25.
11. Charles L. Wilson, ed. *Program Project Mercury Candidate Evaluation Program (Technical Report 59–505)* (Dayton: Wright Air Development

Center, 1959), 1; Mae Mills Link, "Stress Testing," in *Space Medicine in Project Mercury* (Washington, DC: NASA, 1965). Aviation medicine specialists had been discussing the topic informally since the end of World War II, when a group of air force physiologists augmented their numbers with leading German researchers (including Hubertus Strughold, called the "father of space medicine"), invited to the United States after the War.

12. Maura Phillips Mackowski, *Testing the Limits: Aviation Medicine and the Origins of Manned Space Flight* (College Station: Texas A&M University Press, 2006), 167–168.
13. Don Flickinger, "Biomedical Aspects of Space Flight," in Gantz, *Man in Space: The United States Air Force Program for Developing the Spacecraft Crew*, ed. Kenneth Franklin Gantz (New York: Duell, 1959).
14. S. B. Sells and Charles A. Berry, "Human Requirements for Space Flight," in Gantz, *Man in Space*; D. H. Beyer and S. B. Sells, "Selection and Training of Personnel for Space Flight," *Journal of Aviation Medicine* 28 (1957).
15. Sells and Berry, "Human Requirements for Space Flight," 168–170, 172, 174–175.
16. When human spaceflight outgrew the Space Task Group, Gilruth ran its much larger operations at the Manned Spacecraft Center in Houston, Texas, retiring in 1972. John Noble Wilford, "Robert Gilruth, 86, Dies: Was Crucial Player at NASA," *New York Times*, August 18, 2000, C19.
17. "The astronaut selection program within NASA was handicapped at the beginning by the problem of trying to define 'what is an astronaut' and 'what are his duties.' " Stanley C. White, M.D., "Review of Astronaut Selection," June 17, 1963, Flight Crew Operations Directorate, Center Series, Box 3, Johnson Space Center History Collection, University of Houston–Clear Lake.
18. Joseph D. Atkinson and Jay M. Shafritz, *The Real Stuff: A History of NASA's Astronaut Recruitment Program* (New York: Praeger, 1985), 32; Flickinger, "Biomedical Aspects of Space Flight," 46.
19. Philip Kaufman, "The Right Stuff" (The Ladd Company, 1983).
20. Atkinson and Shafritz, *The Real Stuff*, 32–33.
21. "NASA Project A, Announcement 1," December 22, 1958, Folder 013880, NASA Historical Reference Collection, NASA Headquarters, Washington, DC.
22. Atkinson and Shafritz, *The Real Stuff*, 34–35.
23. Lee D. Saegesser, "Report of Telephone Conversation with Dr. T. Keith Glennan," January 24, 1995, Folder 16404, NASA Historical Reference Collection, NASA Headquarters, Washington, DC.
24. Atkinson and Shafritz, *The Real Stuff*, 35–36.
25. George M. Low, "Memorandum to T. Keith Glennan Re: Pilot Selection for Project Mercury," April 23, 1959, Folder 16404, NASA Historical Reference Collection, NASA Headquarters, Washington, DC.
26. Ibid.
27. Atkinson and Shafritz, *The Real Stuff*, 183.
28. Low, "Memorandum to T. Keith Glennan Re: Pilot Selection for Project Mercury."
29. Atkinson and Shafritz, *The Real Stuff*, 182.
30. George E. Ruff and Edwin Z. Levy, "Psychiatric Evaluation of Candidates for Space Flight," *American Journal of Psychiatry* 116 (1959): 385.
31. Mackowski, *Testing the Limits*, 18.

32. Jerome B. Weisner, "Memorandum for Dr. Bundy," March 9, 1961, Flight Crew Operations Directorate, Center Series, Box 3, Johnson Space Center History Collection, University of Houston–Clear Lake.
33. Abe Silverstein, "Memorandum to T. Keith Glennan," January 29, 1959, Folder 16404, NASA Historical Reference Collection, NASA Headquarters, Washington, DC.
34. Ibid.
35. Gilruth, "Transcript #5," 20.
36. Swenson, Grimwood, and Alexander, *This New Ocean, a History of Project Mercury*, 26.
37. Gilruth, "Transcript #5," 20.
38. Mindell, "Human and Machine in the History of Spaceflight," 153–154; Mindell, *Digital Apollo*, 68–73.
39. Milton O. Thompson and Curtis Peebles, *Flying Without Wings: NASA Lifting Bodies and the Birth of the Space Shuttle* (Washington, DC, Smithsonian Institution Press, 1999), 17.
40. Frank Van Riper, *Glenn, the Astronaut Who Would Be President* (New York: Empire Books, 1983), 126.
41. John Catchpole, *Project Mercury: NASA's First Manned Space Programme* (Chichester: Praxis, 2001), 160.
42. Van Riper, *Glenn, the Astronaut Who Would Be President*, 125.
43. Francis French and Colin Burgess, *Into That Silent Sea: Trailblazers of the Space Era, 1961–65* (Lincoln: University of Nebraska Press, 2007), 256.
44. Ed Buckbee and Wally Schirra, *The Real Space Cowboys* (Burlington, Ontario: Apogee Books, 2005); Richard Goldstein, "Walter M. Schirra Jr., Astronaut, Dies at 84," *New York Times*, May 4, 2007; M. Scott Carpenter and et al., *We Seven* (New York: Simon and Schuster, 1962), 78–79.
45. French and Burgess, *Into That Silent Sea*, 205.
46. Neil A. Armstrong, "Oral History Transcript (Dr. Stephen E. Ambrose and Dr. Douglas Brinkley, Interviewers)," NASA Johnson Space Center Oral History Project (Houston, Texas, September 19, 2001), 39.
47. Donald K. Slayton and Michael Cassutt, *Deke! U.S. Manned Space: From Mercury to the Shuttle* (New York: St. Martin's Press, 1994), 67.
48. Atkinson and Shafritz, *The Real Stuff*, 39–40.
49. Colin Burgess, *Selecting the Mercury Seven: The Search for America's First Astronauts* (Chichester: Springer-Praxis, 2011), 38.
50. Walter M. Schirra Jr., "History Transcript (Roy Neal, Interviewer)," NASA Johnson Space Center Oral History Project (San Diego, December 1, 1998), 4.
51. French and Burgess, *Into That Silent Sea*, 135, 204.
52. Atkinson and Shafritz, *The Real Stuff*, 42.
53. French and Burgess, *Into That Silent Sea*, 204–05.
54. Charles L. Wilson, "The Candidate Evaluation Committee," in *Project Mercury Candidate Evaluation Program*, 89; Charles L. Wilson, "Discussion and Recommendations," in *Project Mercury Candidate Evaluation Program*, 99.
55. Patricia A. Santy, *Choosing the Right Stuff: The Psychological Selection of Astronauts and Cosmonauts* (Westport: Praeger, 1994), 34.
56. Robert L. Crippen, "Oral History Transcript (Rebecca Wright, Interviewer)," NASA Johnson Space Center Oral History Project (Houston, Texas, May 26, 2006), 2.
57. David Sington, "In the Shadow of the Moon" (Discovery Films, 2007); Collins, *Carrying the Fire*.

58. See, generally, Burgess, *Selecting the Mercury Seven.*
59. Sington, "In the Shadow of the Moon."
60. Frank Borman, Jim Lovell, and Bill Anders, "John H. Glenn Lecture: An Evening with the Apollo 8 Astronauts," Smithsonian Institution, National Air and Space Museum, November 13, 2008.
61. Al Worden, *Falling to Earth: An Apollo Astronaut's Journey* (Washington, DC: Smithsonian Books, 2011), 146.
62. M. Scott Carpenter and Kris Stoever, *For Spacious Skies: The Uncommon Journey of a Mercury Astronaut* (Orlando: Harcourt, 2002), 183; see, generally, Burgess, *Selecting the Mercury Seven.*
63. Santy, *Choosing the Right Stuff,* 17.
64. National Aeronautics and Space Administration, "Seven to Enter Mercury Training Program (NASA Release No. 59–111)" (Washington, DC: NASA, 1959), 2.
65. See, generally, Burgess, *Selecting the Mercury Seven.*
66. Atkinson and Shafritz, *The Real Stuff,* 43.
67. National Aeronautics and Space Administration, "Mercury Astronaut Selection Fact Sheet" (Washington, DC: NASA, 1959), 2.
68. Joseph P. Allen, "Oral History Transcript (Jennifer Ross-Nazzal, Interviewer)," NASA Johnson Space Center Oral History Project (Houston, Texas, January 28, 2003), 5.
69. Mackowski, *Testing the Limits,* 190–192.
70. C. E. Clauser, "Anthropometric Studies," in *Project Mercury Candidate Evaluation Program,* 24.
71. Wilson, *Program Project Mercury Candidate Evaluation Program,* 89; Atkinson and Shafritz, *The Real Stuff,* 43; Mackowski, *Testing the Limits,* 193.
72. Wilson, *Program Project Mercury Candidate Evaluation Program,* 89.
73. Carpenter and Stoever, *For Spacious Skies,* 192–193.
74. Wilson, *Program Project Mercury Candidate Evaluation Program,* 89.
75. Tom Wolfe, *The Right Stuff* (New York: Farrar, Straus, and Giroux, 1979), 88–95; Slayton and Cassutt, *Deke!*
76. Atkinson and Shafritz, *The Real Stuff,* 45.
77. Wilson, *Program Project Mercury Candidate Evaluation Program.*
78. G. E. Ruff, "Psychological Tests," in *Project Mercury Candidate Evaluation Program.*
79. Ruff and Levy, "Psychiatric Evaluation of Candidates for Space Flight," 389–90.
80. Sheldon J. Korchin and George E. Ruff, "Personality Characteristics of the Mercury Astronauts," in *The Threat of Impending Disaster, Contributions to the Psychology of Stress,* ed. George H. Grosser, Henry Wechsler, and Milton Greenblatt (Cambridge, Massachusetts: MIT Press, 1964), 205.
81. Ruff and Levy, "Psychiatric Evaluation of Candidates for Space Flight," 389.
82. Santy, *Choosing the Right Stuff,* 16.
83. Atkinson and Shafritz, *The Real Stuff,* 30.
84. Ruff and Levy, "Psychiatric Evaluation of Candidates for Space Flight," 389. The notion that astronauts are incapable of anxiety persists in popular culture despite the lack of any empirical support for the claim. In one 2009 article, the author described how a noted professor of psychology was certain that "when the United States sends people up in space, the steely, brave astronauts were low-reactive as infants, and the mission-control people down on the ground, doing the detail work that keeps the craft aloft,

were high-reactive." Robin Marantz Henig, "Understanding the Anxious Mind," *New York Times Magazine*, October 4, 2009, 64. In fact, astronauts fretted constantly about their performance, and were often the most detail-obsessed members of NASA ground teams.

85. French and Burgess, *Into That Silent Sea*, 261.
86. Korchin and Ruff, "Personality Characteristics of the Mercury Astronauts," 204–207.
87. Burgess, *Selecting the Mercury Seven*, 220.
88. Gerard J. De Groot, *Dark Side of the Moon: The Magnificent Madness of the American Lunar Quest* (New York: New York University Press, 2006), 107; Santy, *Choosing the Right Stuff*, 17.
89. Guenter Wendt and Russell Still, *The Unbroken Chain* (Burlington, Ontario: Apogee Books, 2001), 54; French and Burgess, *Into That Silent Sea*, 258; De Groot, *Dark Side of the Moon*, 105–06; Carpenter and Stoever, *For Spacious Skies*, 183.
90. De Groot, *Dark Side of the Moon*, 116.
91. Santy, *Choosing the Right Stuff*, 17.
92. Carpenter and Stoever, *For Spacious Skies*, 189.
93. Wolfe, *The Right Stuff*, 77.
94. Burgess, *Selecting the Mercury Seven*, 257.
95. Wolfe, *The Right Stuff*, 76.
96. Ruff and Levy, "Psychiatric Evaluation of Candidates for Space Flight," 389–90.
97. Ibid., 390.
98. Korchin and Ruff, "Personality Characteristics of the Mercury Astronauts," 202.
99. Santy, *Choosing the Right Stuff*, 18.
100. Mackowski, *Testing the Limits*, 193.
101. Atkinson and Shafritz, *The Real Stuff*, 45–46.
102. Wilson, *Program Project Mercury Candidate Evaluation Program*, 89.
103. Atkinson and Shafritz, *The Real Stuff*, 46; Mackowski, *Testing the Limits*, 193.
104. Korchin and Ruff, "Personality Characteristics of the Mercury Astronauts," 200, 206–207, Carpenter and Stoever, *For Spacious Skies*, 189.
105. Korchin and Ruff, "Personality Characteristics of the Mercury Astronauts," 201. Indeed, the mainstream Christian homogeneity of the early astronauts has been a standing joke in the space program; at the end of Woody Allen's 1971 film *Bananas*, a faux news bulletin flashes across the lower screen announcing that Apollo astronauts have opened "the first all-Protestant cafeteria on the Moon." Woody Allen, "Bananas" (United Artists, 1971).
106. Richard D. Lyons, "Each Astronaut Is an Only Child: Fact Cited as Evidence of Theory of Achievement," *New York Times*, Dec. 24, 1968, 7.
107. Letter from Homer E. Newell to Daniel P. Moynihan, June 10, 1969, Folder 004153, NASA Historical Reference Collection, NASA Headquarters, Washington, DC.
108. Lyons, "Each Astronaut Is an Only Child: Fact Cited as Evidence of Theory of Achievement," 7.
109. Santy, *Choosing the Right Stuff*, 22, 71.
110. National Aeronautics and Space Administration, "Press Conference: Mercury Astronaut Team" (Washington, DC: NASA, 1959), 3.

111. Ibid., 4–5.
112. Wolfe, *The Right Stuff*, 118–120; De Groot, *Dark Side of the Moon*, 107–108.
113. National Aeronautics and Space Administration, "Press Conference: Mercury Astronaut Team," 6, 9–10.
114. Ibid., 15–21.
115. "Men in Space," *New York Times*, April 11, 1959, 20.
116. National Aeronautics and Space Administration, "Press Conference: Mercury Astronaut Team," 26. Washington, DC; April 9, 1959; NASA Historical Reference Collection, NASA Headquarters.
117. John W. Finney, "7 Named as Pilots for Space Flights Scheduled in 1961," *New York Times*, April 10, 1959, 1.
118. James Reston, "Washington: The Sky's No Longer the Limit," *New York Times*, April 12, 1959, E8.
119. While a "farm boy" mythology would eventually emerge around the men in popular culture, this characterization, likely motivated by Wolfe's descriptions of Chuck Yeager in *The Right Stuff*, was not based in fact.
120. E. J. Mclaughlin, "Family Structure of Astronauts" (Houston, Texas: Space Medicine, NASA Manned Space Flight Center, 1969), 3.
121. McCurdy, *Inside NASA* (Baltimore: Johns Hopkins University Press, 1993), 79. See also, for example, Sylvia Doughty Fries, *NASA Engineers and the Age of Apollo* (Washington, DC: NASA, 1992).
122. French and Burgess, *Into That Silent Sea*, 41–44.
123. O'Leary, *The Making of an Ex-Astronaut* (Boston: Houghton Mifflin, 1970), 164.
124. French and Burgess, *Into That Silent Sea*, 44–45.
125. Howard Muson, "Comedown from the Moon: What Has Happened to the Astronauts," *New York Times Magazine*, December 3, 1972, 135.
126. McCurdy, *Space and the American Imagination* (Washington, DC: Smithsonian Institution Press, 1997), 88.
127. James L. Schefter, *The Race: The Uncensored Story of How America Beat Russia to the Moon* (New York: Doubleday, 1999), 87; De Groot, *Dark Side of the Moon*, 112.
128. Walter Cunningham, *The All-American Boys (Revised Edition)* (New York: ibooks, 2003).
129. Collins, *Carrying the Fire*, 25.
130. Society of Experimental Test Pilots, *History of the First 20 Years*, 73, 118.
131. See, for example, Carpenter and Stoever, *For Spacious Skies*; Wally Schirra and Richard N. Billings, *Schirra's Space* (Boston: Quinlan Press, 1988).
132. Kenneth Keniston, *Youth and Dissent: The Rise of a New Opposition* (New York: Harcourt Brace Jovanovich, 1971), 110.
133. Buckbee and Schirra, *The Real Space Cowboys*; Goldstein, "Walter M. Schirra Jr., Astronaut, Dies at 84."
134. Francis French and Colin Burgess, *In the Shadow of the Moon: A Challenging Journey to Tranquility, 1965–69* (Lincoln: University of Nebraska Press, 2007), 82.
135. James R. Hansen, *First Man: The Life of Neil A. Armstrong* (New York: Simon & Schuster, 2005), 137.
136. Robert Wells, *What Does an Astronaut Do?* (New York: Dodd, 1961), 8–9.
137. Ruff and Levy, "Psychiatric Evaluation of Candidates for Space Flight," 385.
138. For example, Slayton and Cassutt, *Deke!*

139. Society of Experimental Test Pilots., *History of the First 20 Years*, 62; De Groot, *Dark Side of the Moon*, 112.
140. Schefter, *The Race*, 86–87.
141. Mindell, *Digital Apollo*, 80; Gladwin Hill, "Test Pilots Get Some Good News," *New York Times*, October 9, 1959.
142. Catchpole, *Project Mercury*, 160.
143. Atkinson and Shafritz, *The Real Stuff*, 28.
144. Hubert M. Drake, Donald R. Bellman, and Joseph A. Walker, "Research Memorandum: Orbital Problems of Manned Orbital Vehicles (April 12, 1958)," in, Robert, Godwin ed., *Dyna-Soar: Hypersonic Strategic Weapons System* (Burlington, Ontario: Apogee Books, 2003), 93.
145. Gilruth, "Transcript #4," 35.
146. Carpenter, *We Seven*, 97–98.
147. Ibid., 90, 101.
148. Ibid., 91.
149. Schefter, *The Race*, 88.
150. Robert Gilruth, "Transcript #6," Oral History Transcripts (David DeVorkin, John Mauer, Interviewers) (Washington, DC: Smithsonian Institution, National Air and Space Museum, March 2, 1987), 83–84.
151. Carpenter, *We Seven*, 108.
152. Schefter, *The Race*, 88–91.
153. Betty Grissom and Henry Still, *Starfall* (New York: Crowell, 1974), 149.
154. Asif A. Siddiqi, *Challenge to Apollo: The Soviet Union and the Space Race, 1945–1974* (Washington, DC: NASA, 2000), 173.
155. French and Burgess, *Into That Silent Sea*, 80.
156. Donna Jeanne Haraway, *Primate Visions: Gender, Race, and Nature in the World of Modern Science* (New York: Routledge, 1989), 138.
157. Slayton and Cassutt, *Deke!*, 67.
158. Homer A. Boushey, "Blueprints for Space," in Gantz, *Man in Space*, 247; Haughty, "Human Performance in Space," 85.
159. Collins, *Carrying the Fire*, 37.
160. J. E. Tomayko, *Computers in Space: Journeys with NASA* (Indianapolis: Alpha Books, 1994), 25.
161. Norman Mailer, *Of a Fire on the Moon* (Boston: Little, Brown, 1970), 28.
162. French and Burgess, *Into That Silent Sea*, 257.
163. Korchin and Ruff, "Personality Characteristics of the Mercury Astronauts," 204.
164. National Aeronautics and Space Administration, "Press Conference: Mercury Astronaut Team," 3.
165. "George M. Low," in *Before This Decade Is out: Personal Reflections on the Apollo Program*, ed. Glen E. Swanson (Washington, DC: NASA, 1999), 337; Carpenter, *We Seven*, 229–230.
166. Ibid. "We were left out of that decision-making," Carpenter later recalled. M. Scott Carpenter, "Oral History Transcript (Michelle Kelly, Interviewer)," NASA Johnson Space Center Oral History Project (Houston, Texas, March 30, 1998), 9.
167. George E. Ruff and Sheldon J. Korchin, "Psychological Responses of the Mercury Astronauts to Stress," in *The Threat of Impending Disaster, Contributions to the Psychology of Stress*, ed. George H. Grosser, Henry Wechsler, and Milton Greenblatt (Cambridge, Massachusetts: MIT Press, 1964), 216.

168. Schirra and Billings, *Schirra's Space*, 71–72.
169. Ibid.
170. Michael J. Neufeld, *Von Braun: Dreamer of Space, Engineer of War* (New York: A.A. Knopf, 2007).
171. Slava Gerovitch, " 'New Soviet Man' inside the Machine: Human Engineering, Spacecraft Design, and the Construction of Communism," *Osiris* 22 (2007): 141, 147–148.
172. For example, Siddiqi, *Challenge to Apollo*, 271–283; See, for example, Jamie Doran, *Starman: The Truth Behind the Legend of Yuri Gagarin* (London: Bloomsbury, 1998), 82.
173. Gilruth, "Transcript #6," 18, 20.
174. Carpenter, *We Seven*, 220.
175. Gilruth, *Transcript #4*, 29.
176. Indeed, as Apollo 8 astronaut William Anders later noted, "as fighter pilots, it's much better to die than screw up." Borman, Lovell, and Anders, "John H. Glenn Lecture: An Evening with the Apollo 8 Astronauts."
177. French and Burgess, *Into That Silent Sea*, 235.
178. Ibid., 284–285, Neal Thompson, *Light This Candle: The Life and Times of Alan Shepard, America's First Spaceman* (New York: Crown Publishers, 2004), 290–291.
179. "Astronauts Push for a 7th Flight," *New York Times*, May 22, 1963.
180. Schirra, "History Transcript (Roy Neal, Interviewer)," 5.
181. Wendt and Still, *The Unbroken Chain*, 53–54.
182. Schirra and Billings, *Schirra's Space*, 77.
183. Muson, "Comedown from the Moon," 136–137. Shepard never acknowledged a subordinate role to Slayton, and the two often acted in concert, especially in assigning crews to flights.
184. Neufeld, *Von Braun: Dreamer of Space, Engineer of War*, 357–358; David J. Shayler, *NASA's Scientist-Astronauts* (New York: Springer, 2007), 14.
185. Wernher Von Braun, "Wernher Von Braun to the Vice President of the United States, April 29, 1961," in *Exploring the Unknown: Selected Documents in the History of the U.S. Civil Space Program*, ed. John M. Logsdon (Washington, DC: NASA, 1995).
186. Lyndon B. Johnson, "Evaluation of Space Program" (April 28, 1961).
187. Cunningham, *The All-American Boys (Revised Edition)*, 285.

2 "Deke's Boys"

1. Michael Collins, *Carrying the Fire: An Astronaut's Journeys* (New York: Farrar, 1974), 179.
2. Ibid., 14.
3. Francis French and Colin Burgess, *In the Shadow of the Moon: A Challenging Journey to Tranquility, 1965–69* (Lincoln: University of Nebraska Press, 2007), 65, 113–114.
4. National Aeronautics and Space Administration, "Average Astronaut (MSC 65–22)," *Manned Spacecraft Center Press Release*, February 10, 1965, 1.
5. Collins, *Carrying the Fire*, 60; French and Burgess, *In the Shadow of the Moon*, 58–59, 63–65.
6. "The problem of selection encountered during the second cycle was considerably different and, in the main much easier." Staley C. White, M.D.,

"Review of Astronaut Selection," June 17, 1963, Flight Crew Operations Directorate, Center Series, Box 3, Johnson Space Center History Collection, University of Houston-Clear Lake.

7. Collins, *Carrying the Fire*, 26–27, 33.
8. Walter Cunningham, *The All-American Boys (Revised Edition)* (New York: ibooks, 2003), 98; Collins, *Carrying the Fire*, 99.
9. Norman Mailer, *Of a Fire on the Moon* (Boston: Little, Brown, 1970), 8–9.
10. French and Burgess, *In the Shadow of the Moon*, 125.
11. Collins, *Carrying the Fire*, 51, 176–177.
12. French and Burgess, *In the Shadow of the Moon*, 142 [quoting Grissom at a press conference], 205.
13. David A. Mindell, *Digital Apollo: Human and Machine in Spaceflight* (Cambridge, Massachusetts: MIT Press, 2008), 158, 168.
14. Alan L. Bean, "Oral History Transcript (Michelle Kelly, Interviewer)," NASA Johnson Space Center Oral History Project (Houston, Texas, June 23, 1998), 7–9.
15. For example, "Use Of Astronauts In Quality Assurance Visits," March 3, 1964, Box 064–24, Apollo Series, Johnson Space Center History Collection, University of Houston–Clear Lake; Mindell, *Digital Apollo*, 155.
16. Cunningham, *The All-American Boys (Revised Edition)*, 43–44.
17. Robert L. Crippen, "Oral History Transcript (Rebecca Wright, Interviewer)," NASA Johnson Space Center Oral History Project (Houston, Texas, May 26, 2006), 4.
18. Tom Wolfe, *The Right Stuff* (New York: Farrar, Straus, and Giroux, 1979), 143.
19. Borman, Lovell, and Anders, "John H. Glenn Lecture: An Evening with the Apollo 8 Astronauts." See, generally, Kristen Starr, "NASA's Hidden Power: NACA/NASA Public Relations and the Cold War, 1945–1967" (PhD diss., Auburn University, 2008).
20. George M. Low, "Handwritten Annotation Upon Memorandum from John P. Donnelly to George M. Low," November 20, 1972, Folder 1, Box 50, George M. Low Papers, Archives and Special Collections, Rensselaer Polytechnic Institute, Troy, New York.
21. Cunningham, *The All-American Boys (Revised Edition)*, 189, 192.
22. William E. Howard, "USIA Editors Scooped on Flight Stories," *The Birmingham News*, May 29, 1963, 16.
23. National Aeronautics and Space Administration, "NASA Policy Concerning Mercury Astronauts (NASA News Release)," May 11, 1959, NASA Historical Reference Collection, NASA Headquarters, Washington, DC.
24. Robert Sherrod, "The Selling of the Astronauts," *Columbia Journalism Review*, May/June 1973, 17; "Agreement between Malcom Scott Carpenter, et al. and C. Leo Deorsey," May 28, 1959, Folder 008937, NASA Historical Reference Collection, NASA Headquarters, Washington, DC.
25. Sherrod, "The Selling of the Astronauts," 18–19.
26. Cunningham, *The All-American Boys (Revised Edition)*, 194.
27. Sherrod, "The Selling of the Astronauts," 20.
28. "Memorandum from Walter L. Lingle Jr. For the Deputy Administrator Re: Ad Hoc Committee on Astronauts; Life Contract," July 26, 1962, Folder 008937, NASA Historical Reference Collection, NASA Headquarters, Washington, DC.
29. The contracts' benefits, though, were nearly outweighed by the criticism they generated from NASA managers, governmental officials, the public,

and the press, for whom the agreement represented a form of "crass commercialism" that cheapened the space program. Walter L. Lingle Jr., "Memorandum for the Deputy Administrator Re: Ad Hoc Committee on Astronauts; Life Contract," July 26, 1962, Folder 008937, NASA Historical Reference Collection, NASA Headquarters, Washington, DC. Editors of competing publications were particularly furious, concerned by the blurry boundary between public and private aspects of the astronauts' stories and afraid that NASA had effectively granted a few publications exclusive rights to major public figures. Sherrod, "The Selling of the Astronauts," 18. Among those "cut out of the picture," in fact, was the federal government's own United States Information Agency, which found its efforts to interview the astronauts blocked by the life contract. William E. Howard, "USIA Editors Scooped on Flight Stories," *The Birmingham News*, May 29,1963, 16.

30. James E. Webb, "Letter to Robert Sherrod," August 20, 1973, Folder 004157, NASA Historical Reference Collection, NASA Headquarters, Washington, DC.
31. Cunningham, *The All-American Boys (Revised Edition)*, 193.
32. French and Burgess, *In the Shadow of the Moon*, xii.
33. Cunningham, *The All-American Boys (Revised Edition)*, 192; Sherrod, "The Selling of the Astronauts," 17–24.
34. Sherrod, "The Selling of the Astronauts," 23.
35. Cunningham, *The All-American Boys (Revised Edition)*, 197–199.
36. "Interrogatories of Alan B. Shepard Jr." (National Aeronautics and Space Administration, 1972).
37. David J. Shayler, *NASA's Scientist-Astronauts* (New York: Springer, 2007), 195.
38. Allen, Oral History Transcript (Jennifer Ross-Nazzal, Interviewer), 7.
39. "Investigation Report, Admiral Alan B. Shepard Jr.," August 29, 1972, Folder 008943, NASA Historical Reference Collection, NASA Headquarters, Washington, D.C.
40. French and Burgess, *In the Shadow of the Moon*, xi.
41. George M. Low, "Letter to Clinton P. Anderson," July 10, 1972, p. 1, Folder 004156, NASA Historical Reference Collection, NASA Headquarters, Washington, DC.
42. Cunningham, *The All-American Boys (Revised Edition)*, 198.
43. Korchin and Ruff, "Personality Characteristics of the Mercury Astronauts," in *The Threat of Impending Disaster, Contributions to the Psychology of Stress*, ed. George H. Grosser, Henry Wechsler, and Milton Greenblatt (Cambridge, Massachusetts: MIT Press, 1964), 202.
44. For example, Merritt Roe Smith, *Harpers Ferry Armory and the New Technology: The Challenge of Change* (Ithaca, New York: Cornell University Press, 1977).
45. Korchin and Ruff, "Personality Characteristics of the Mercury Astronauts," 202.
46. Cunningham, *The All-American Boys (Revised Edition)*, 210, 215, 226.
47. Wolfe, *The Right Stuff*, 27–28, 329. While NASA prohibited pilots from flying jets less than 12 hours after consuming alcohol (the so-called "bottle-to-throttle" rule), excessive off-duty drinking at Cape Kennedy often extended well into to the morning of space launches, for which no such rule applied. In 2007, NASA released unsubstantiated reports that two astronauts may have actually flown in space while so intoxicated that their performance

was impaired, despite cautions from flight surgeons. For example, Warren E. Leary, "NASA Opening an Inquiry into Drunken-Flying Reports," *New York Times*, July 28, 2007, A11; John Schwartz, "Astronauts Have Flown While Drunk, NASA Finds," *New York Times*, July 27, 2007, A16. NASA officially disproved this report but it remains controversial.

48. National Aeronautics and Space Administration, "Press Conference: Mercury Astronaut Team," 4; Howard E. McCurdy, *Space and the American Imagination* (Washington, DC: Smithsonian Institution Press, 1997), 88–91; Gerard J. De Groot, *Dark Side of the Moon: The Magnificent Madness of the American Lunar Quest* (New York: New York University Press, 2006), 106.
49. French and Burgess, *In the Shadow of the Moon*, 198.
50. Cunningham, *The All-American Boys (Revised Edition)*, 217.
51. Borman, Lovell, and Anders, "John H. Glenn Lecture: An Evening with the Apollo 8 Astronauts."
52. Howard Muson, "Comedown from the Moon: What Has Happened to the Astronauts," *New York Times Magazine*, December 3, 1972, 136.
53. Joseph D. Atkinson and Jay M. Shafritz, *The Real Stuff: A History of NASA's Astronaut Recruitment Program* (New York: Praeger, 1985), 30.
54. Cunningham, *The All-American Boys (Revised Edition)*, 213–214.
55. Shayler, *NASA's Scientist-Astronauts*, 83–85.
56. Cunningham, *The All-American Boys (Revised Edition)*, 213–214.
57. French and Burgess, *In the Shadow of the Moon*, 186, 204, 206–208, 219, 233, 235–236 [quoting authors' interviews of Schirra and Cunningham].
58. Cunningham, *The All-American Boys (Revised Edition)*, 226.
59. Ibid., 216–217.
60. Ibid., 212.
61. Ibid., 98–99, 189.
62. Shayler, *NASA's Scientist-Astronauts*, 247.
63. "Memorandum from W. F. Boone to Webb and Seamens Re: Commendatory Promotions for Astronauts," August 5, 1965, NASA Historical Reference Collection, NASA Headquarters, Washington, DC, Folder 008949.
64. Joseph A. Califano Jr., "Memorandum to Secretary of Defense; Administrator, National Aeronautics and Space Administration," August 17, 1965, Folder 008949, NASA Historical Reference Collection, NASA Headquarters, Washington, DC.
65. Samuel Keller, "Memorandum for the Record Re: Grade Level for Civilian Astronauts," December 16, 1975, Folder 008949 NASA Historical Reference Collection, NASA Headquarters, Washington, DC.
66. J. M. Poindexter, "Memorandum to Caspar W. Weinberger and William R. Graham," December 16, 1985, Folder 008949, NASA Historical Reference Collection, NASA Headquarters, Washington, DC.
67. James C. Fletcher, "Letter to John C. Stennis," June 6, 1972, Folder 004156, NASA Historical Reference Collection, NASA Headquarters, Washington, DC.
68. Cunningham, *The All-American Boys (Revised Edition)*, 99.
69. Shayler, *NASA's Scientist-Astronauts*, 244.
70. See, generally, Edwards, *Contested Terrain* (New York: Basic Books, 1979).
71. Robert Gilruth, "Transcript #6," Oral History Transcripts (David DeVorkin, John Mauer, Interviewers) (Washington, DC.: Smithsonian Institution, National Air and Space Museum, March 2, 1987), 85.
72. Collins, *Carrying the Fire*, 313.

73. "George M. Low," in *Before This Decade Is out: Personal Reflections on the Apollo Program*, ed. Glen E. Swanson (Washington, DC: NASA, 1999), 337–338.
74. Eugene Cernan and Don Davis, *The Last Man on the Moon: Astronaut Eugene Cernan and America's Race in Space* (New York: St. Martin's Press, 1999), 67.
75. David J. Shayler, *Skylab: America's Space Station* (Chichester, UK: Praxis, 2001), 123. This question often arises in connection with the crew selected for the first attempted lunar landing, Apollo 11—Neil Armstrong, Buzz Aldrin, and Michael Collins. Slayton has insisted that any of NASA's astronauts could have successfully flown that mission and that it was merely the crew's turn in the rotation. Apollo 11's crew, though, was composed entirely of veteran astronauts—a rarity—known for their unique skills and abilities.
76. Gilruth, "Transcript #6," 85.
77. Cernan and Davis, *The Last Man on the Moon*, 67.
78. Gilruth, "Transcript #6," 85.
79. Donald K. Slayton and Michael Cassutt, *Deke! U.S. Manned Space: From Mercury to the Shuttle* (New York: St. Martin's Press, 1994), 136; Hansen, *First Man* (New York: Simon & Schuster, 2005), 231.
80. Hansen, *First Man*, 231.
81. Slayton and Cassutt, *Deke!*, 136.
82. Ibid.
83. Molly Ivins, "Ed Who?," *New York Times*, June 30, 1974, 15.
84. French and Burgess, *In the Shadow of the Moon*, 185 (quoting Cunningham).
85. Shayler, *Skylab*, 105.
86. Cunningham, *The All-American Boys (Revised Edition)*, 48.
87. Slayton and Cassutt, *Deke!*, 145.
88. Ibid., 143.
89. Cunningham, *The All-American Boys (Revised Edition).*
90. Slayton and Cassutt, *Deke!*, 134; French and Burgess, *In the Shadow of the Moon*, 196.
91. Cunningham, *The All-American Boys (Revised Edition)*, 48; Slayton and Cassutt, *Deke!*, 134.
92. French and Burgess, *In the Shadow of the Moon*, 293.
93. Ibid.; Shayler, *Skylab*, 304–305.
94. Cunningham, *The All-American Boys (Revised Edition)*, 39.
95. Shayler, *Skylab*, 305; French and Burgess, *In the Shadow of the Moon*, 185.
96. Cunningham, *The All-American Boys (Revised Edition).*
97. French and Burgess, *In the Shadow of the Moon*, 64.
98. The astronauts' accumulation of influence bears a striking resemblance to the manner in which scientists acquire and spend "credit" within the laboratory setting. Bruno Latour and Steve Woolgar, *Laboratory Life: The Construction of Scientific Facts* (Princeton: Princeton University Press, 1986); Sharon Traweek, *Beamtimes and Lifetimes: The World of High Energy Physicists* (Cambridge, Massachusetts: Harvard University Press, 1988). Like the particle physicists Traweek describes, senior astronauts could expend their influence to assign themselves to coveted missions, but these raw "power plays" offended other astronauts. Cunningham, *The All-American Boys (Revised Edition)*, 330.
99. Thomas P. Stafford and Michael Cassutt, *We Have Capture: Tom Stafford and the Space Race* (Washington, DC: Smithsonian Institution Press, 2002),

142–144; Shayler, *Skylab*, 122; Cernan and Davis, *The Last Man on the Moon*, 238; Slayton and Cassutt, *Deke!*, 237.

100. Cernan and Davis, *The Last Man on the Moon*, 263.
101. Cunningham, *The All-American Boys (Revised Edition)*, 112.
102. Ibid., 243; French and Burgess, *In the Shadow of the Moon*, 361; Shayler, *Skylab*, 118–120.
103. Slayton and Cassutt, *Deke!*; Stafford and Cassutt, *We Have Capture*.
104. French and Burgess, *In the Shadow of the Moon*, 319.
105. Cunningham, *The All-American Boys (Revised Edition)*, 43.
106. French and Burgess, *In the Shadow of the Moon*, 123, 125, 126.
107. Ibid., 199, 294–295.
108. Alan L. Bean, "Oral History Transcript (Michelle Kelly, Interviewer)," 10.
109. Ibid., 5, 15.
110. Cunningham, *The All-American Boys (Revised Edition)*, 47.
111. Ibid., 134.
112. Ruff and Korchin, "Psychological Responses of the Mercury Astronauts to Stress," in *The Threat of Impending Disaster, Contributions to the Psychology of Stress,* ed. George H. Grosser, Henry Wechsler, and Milton Greenblatt (Cambridge: MIT Press, 1964), 216.
113. Ibid., 10.
114. Alan L. Bean, "Oral History Transcript (Michelle Kelly, Interviewer)," 20.
115. Mindell, *Digital Apollo*, 208.
116. Alan L. Bean, "Oral History Transcript (Michelle Kelly, Interviewer)," 12.
117. Crippen, "Oral History Transcript (Rebecca Wright, Interviewer)," 22.
118. Mindell, *Digital Apollo*, 63.
119. French and Burgess, *In the Shadow of the Moon*, 70.
120. Collins, *Carrying the Fire*, 197–198.
121. Flight Crew Support Division Manned Operations Branch, "Apollo 11 Technical Crew Debriefing (U), Vol. 2" (Houston: NASA, 1969), 24.50.
122. Ibid., 24.53.
123. Carpenter, *We Seven* (New York: Simon and Schuster, 1962), 281.
124. Collins, *Carrying the Fire*, 355.
125. Edward G. Gibson, "Oral History Transcript (Carol Butler, Interviewer)," NASA Johnson Space Center Oral History Project (Houston, Texas, December 1, 2000), 53–55.
126. Wolfgang Schivelbusch, *The Railway Journey: Trains and Travel in the 19th Century* (New York: Urizen Books, 1979).
127. Courtney G. Brooks, James M. Grimwood, and Loyd S. Swenson, *Chariots for Apollo: A History of Manned Lunar Spacecraft* (Washington, DC: NASA, 1979), 76.
128. Collins, *Carrying the Fire*, 215, 223; French and Burgess, *In the Shadow of the Moon*, 109–110.
129. Collins, *Carrying the Fire*, 215.
130. French and Burgess, *In the Shadow of the Moon*, 306–307.
131. Ibid., 304.
132. Manned Operations Branch, "Apollo 11 Technical Crew Debriefing (U), Vol. 2," 25.12–13.
133. French and Burgess, *In the Shadow of the Moon*, 306–307.
134. Crippen, "Oral History Transcript (Rebecca Wright, Interviewer)," 28.
135. French and Burgess, *In the Shadow of the Moon*, 214.
136. Ibid., 69.

137. Ibid., 76.
138. De Groot, *Dark Side of the Moon*, 226.
139. Flight Crew Support Division Manned Operations Branch, "Apollo 11 Technical Crew Debriefing (U), Vol. 1" (Houston: NASA, 1969), 25.22.
140. Catchpole, *Project Mercury: NASA's First Manned Space Programme* (Chichester: Praxis, 2001), 471.
141. Mindell, *Digital Apollo*, 137; Mary Louise Morse and Jean Kernahan Bays, *Apollo Spacecraft: A Chronology, Volume II* (Washington, DC: National Aeronautics and Space Administration, 1973), 99.
142. Mindell, *Digital Apollo*, 138.
143. Arnauld E. Nicogossian, Carolyn Leach Huntoon, and Sam L. Pool, *Space Physiology and Medicine* (Philadelphia: Lea & Febiger, 1994), 436.
144. Ibid., 121; Mark Wade, "Apollo 7," Encyclopedia Astronautica, http://www.astronautix.com/flights/apollo7.htm (accessed April 4, 2010).
145. Ivins, "Ed Who?," 12.
146. Collins, *Carrying the Fire*, 215.
147. French and Burgess, *In the Shadow of the Moon*, 306.
148. Manned Operations Branch, "Apollo 11 Technical Crew Debriefing (U), Vol. 2," 25.10.
149. French and Burgess, *In the Shadow of the Moon*, 215.
150. Mindell, *Digital Apollo*, 105.
151. Ibid., 165.
152. Ibid., ch. 4.
153. Shayler, *NASA's Scientist-Astronauts*, 190.
154. Mindell, *Digital Apollo*, 86.
155. Collins, *Carrying the Fire*, 37; French and Burgess, *In the Shadow of the Moon*, 28, 112.
156. Ibid., 72.
157. Ibid., 213.
158. Mindell, *Digital Apollo*, 215.
159. Harry Braverman, *Labor and Monopoly Capital: The Degradation of Work in the Twentieth Century* (New York: Monthly Review Press, 1975); David F. Noble, *Forces of Production: A Social History of Industrial Automation* (New York: Knopf, 1984).
160. French and Burgess, *In the Shadow of the Moon*, 305.
161. Manned Operations Branch, "Apollo 11 Technical Crew Debriefing (U), Vol. 2," 25.12.
162. Collins, *Carrying the Fire*, 311.
163. See, Matthew H. Hersch, "Checklist: The Secret Life of Apollo's 'Fourth Crewmember'," in *Space Travel & Culture: From Apollo to Space Tourism*, ed. David Bell and Martin Parker (Oxford, UK: Blackwell, 2009).
164. National Aeronautics and Space Administration, *Apollo 11 Technical Air-to-Ground Voice Transcription (GOSS NET 1)* (Houston: Manned Spacecraft Center, 1969), 599.
165. Ibid., 528–529.
166. French and Burgess, *In the Shadow of the Moon*, 11.
167. Ibid., 32–33.
168. C. Gordon Fullerton, "Oral History Transcript (Rebecca Wright, Interviewer)," NASA Johnson Space Center Oral History Project (NASA Dryden Flight Research Center, California, May 6, 2002), 27.
169. French and Burgess, *In the Shadow of the Moon*, 48.

170. Ibid., 48–49.
171. Ruff and Korchin, "Psychological Responses of the Mercury Astronauts to Stress," 217, 219.
172. Fullerton, "Oral History Transcript (Rebecca Wright, Interviewer)," 36.
173. Ruff and Korchin, "Psychological Responses of the Mercury Astronauts to Stress," 218. While few astronauts ever admitted to being afraid while in orbit, space vehicles presented so many different potential failure modes that Michael Collins, one of the more openly self-reflective of the astronauts, later admitted to being "mildly worried all the time" during his Apollo 11 flight. Sington, "In the Shadow of the Moon." Motion picture (Discovery Films, 2007).
174. See, also, Thomas Mallon, "Moon Walker: How Neil Armstrong Brought the Space Program Down to Earth," *The New Yorker*, October 3, 2005; Hansen, *First Man*.
175. Mailer, *Of a Fire on the Moon*, 334.
176. Korchin and Ruff, "Personality Characteristics of the Mercury Astronauts," 204.
177. De Groot, *Dark Side of the Moon*, 107–108.
178. Andrew Smith, *Moondust: In Search of the Men Who Fell to Earth* (New York: Fourth Estate, 2005).
179. Ibid., 218.
180. Korchin and Ruff, "Personality Characteristics of the Mercury Astronauts," 204.
181. French and Burgess, *In the Shadow of the Moon*, 70–71.
182. Ibid., 86–87.
183. Sington, "In the Shadow of the Moon."
184. McCurdy, *Space and the American Imagination*, 84.
185. Ibid., 92.
186. Cunningham, *The All-American Boys (Revised Edition)*, 103.
187. Hansen, *First Man*, 331–332.
188. Collins, *Carrying the Fire*, 215.
189. French and Burgess, *In the Shadow of the Moon*, 315–316; Mindell, *Digital Apollo*, 179.
190. Sington, "In the Shadow of the Moon"; Manned Operations Branch, "Apollo 11 Technical Crew Debriefing (U), Vol. 1," 9.2; Mindell, *Digital Apollo*, 231.
191. Grissom had failed to seal the oxygen inlet on his suit before disembarking, causing it to fill with seawater and nearly drowning him. Fortunately, the neck of Grissom's suit was sealed against water penetration by a rubber barrier Schirra had recently added to the suit; an innovation credited with saving Grissom's life. Even worse, during the investigation, Grissom admitted to needlessly weighing himself down with coins, dollar bills, and other souvenirs, and noted that immediately before the hatch explosion he had contemplated retrieving his survival knife as a keepsake. Even so, when a helicopter dropped a padded rescue strap to pull him out of the water, Grissom, exhausted, climbed into it backwards. The capsule, filled with too much water to be lifted by helicopter, sank rapidly. Grissom, once safely ensconced in the helicopter, grabbed a life preserver and spent his short flight to an awaiting aircraft carrier fastening it, apparently startled by the incident. French and Burgess, *Into That Silent Sea*, 84–88.
192. French and Burgess, *In the Shadow of the Moon*, 1–2, 4.
193. "George M. Low," 338.
194. French and Burgess, *Into That Silent Sea*, 284–285.
195. Ibid., 158–159.

196. Cunningham, *The All-American Boys (Revised Edition)*, 96.
197. Slayton and Cassutt, *Deke!*, 114.
198. French and Burgess, *Into That Silent Sea*, 160, Ruff and Korchin, "Psychological Responses of the Mercury Astronauts to Stress."
199. Kraft later described this event in a chapter of his 2001 memoir entitled, "The Man Malfunctioned." Christopher C. Kraft, *Flight: My Life in Mission Control* (New York: Dutton, 2001); Slayton and Cassutt, *Deke!*, 114; Catchpole, *Project Mercury*, 350–351; Wolfe, *The Right Stuff*, 369–370.
200. Cunningham, *The All-American Boys (Revised Edition)*, 96–97.
201. French and Burgess, *Into That Silent Sea*, 42 [quoting authors' 2003 interview of Dee O'Hara], 261, 266–268.
202. L. Gordon Cooper Jr., "Oral History Transcript (Roy Neal, Interviewer)" (Pasadena, California, May 21, 1998), 13–15.
203. Shayler, *NASA's Scientist-Astronauts*, 246.
204. French and Burgess, *In the Shadow of the Moon*, 50.
205. Astronauts may be analogizes to other groups of workers whose ethnic or religious differences interfered with efforts to organize and bargain collectively in the workplace. See, for example, Bruce Nelson, *Divided We Stand: American Workers and the Struggle for Black Equality*, *Politics and Society in Twentieth-Century America* (Princeton: Princeton University Press, 2001).
206. French and Burgess, *In the Shadow of the Moon*, 37, 227, 233.

3 Scientists in Space

1. Joseph D. Atkinson and Jay M. Shafritz, *The Real Stuff: A History of NASA's Astronaut Recruitment Program* (New York: Praeger, 1985), 64–65, 82; Robert Gilruth, "Transcript #4 (Martin Collins, David DeVorkin, Interviewers)," Oral History Transcripts (Washington, DC: Smithsonian Institution, National Air and Space Museum, October 2, 1986), 30.
2. George M. Low, "Letter to Gordon Allett," November 30, 1970, Folder 004154, NASA Historical Reference Collection, NASA Headquarters, Washington, DC.
3. Jules Verne, *From the Earth to the Moon: and, a Trip around It* (Philadelphia: Lippincott, 1950).
4. Occasionally joining the "real" crew were the professor's comely female assistant, a monkey, a child, or an imbecile, to provide comic relief, dramatic tension, a romantic partner for the ship's pilot, or a surrogate for the audience to whom plot points could be explained. Byron Haskin, "Robinson Crusoe on Mars" (Paramount Pictures Corp., 1964); Kurt Neumann, "Rocketship X-M" (Lippert Pictures, 1950).
5. David A. Mindell, *Digital Apollo: Human and Machine in Spaceflight* (Cambridge, Massachusetts: MIT Press, 2008), 67.
6. Wernher Von Braun and Cornelius Ryan, *Conquest of the Moon* (New York: Viking Press, 1953), 36–38. "The captain of the ship, on being told by his navigator that the vehicle is off course, can make the desired change by inserting a previously prepared tape into the automatic pilot." Von Braun and Ryan, *Conquest of the Moon*, 48.
7. Brian O'Leary, "Topics: Science or Stunts on the Moon?," *New York Times*, April 25, 1970, 18; Ralph E. Lapp, "Send Computers, Not Men, into Deep Space," *New York Times*, February 2, 1969, 35–36.

8. Lapp, "Send Computers, Not Men, into Deep Space," 35–36.
9. Atkinson and Shafritz, *The Real Stuff*, 57–58.
10. Robert Gilruth, "Transcript #6," Oral History Transcripts (David DeVorkin, John Mauer, Interviewers) (Washington, DC.: Smithsonian Institution, National Air and Space Museum, March 2, 1987), 17.
11. See, for example, Walter A. McDougall, *The Heavens and the Earth*, (Baltimore: Johns Hopkins University Press, 1997).
12. Lyndon B. Johnson, "Evaluation of Space Program." in *Exploring the Unknown: Selected Documents in the History of the U.S. Civil Space Program*, ed. John M. Logsdon (Washington, DC: NASA, 1995), 427–429.
13. McDougall, *The Heavens and the Earth*, 413.
14. Atkinson and Shafritz, *The Real Stuff*, 184.
15. Space Task Group, "The Post-Apollo Space Program: Directions for the Future," in *Exploring the Unknown: Selected Documents in the History of the U.S. Civil Space Program*, ed. John M. Logsdon and Linda J. Lear (Washington, DC: NASA, 1995), 526–527.
16. Atkinson and Shafritz, *The Real Stuff*, 65.
17. Ibid., 61–66.
18. O'Leary, "Topics: Science or Stunts on the Moon?"
19. Atkinson and Shafritz, *The Real Stuff*, 64.
20. Sylvia Doughty Fries, *NASA Engineers and the Age of Apollo* (Washington, DC: NASA, 1992), 126,130; Atkinson and Shafritz, *The Real Stuff*, 65.
21. Tom Wolfe, *The Right Stuff* (New York: Farrar, Straus, and Giroux, 1979).
22. Walter Cunningham, *The All-American Boys (Revised Edition)* (New York: ibooks, 2003), 285.
23. Generally, Allan A. Needell, *Science, Cold War and the American State: Lloyd V. Berkner and the Balance of Professional Ideals* (Amsterdam: Harwood Academic Publishers, 2000).
24. Atkinson and Shafritz, *The Real Stuff*, 68–70.
25. David J. Shayler, *NASA's Scientist-Astronauts* (New York: Springer, 2007), 19; Donald K. Slayton and Michael Cassutt, *Deke! U.S. Manned Space: From Mercury to the Shuttle* (New York: St. Martin's Press, 1994).
26. Shayler, *NASA's Scientist-Astronauts*, 29–30.
27. Slayton and Cassutt, *Deke!*, 143.
28. William Lee, "Memorandum to Joseph Shea, May 14, 1963," in *The Real Stuff: A History of NASA's Astronaut Recruitment Program*, ed. Joseph D. Atkinson and Jay M. Shafritz (New York: Praeger, 1985), 71.
29. Ibid.
30. Shayler, *NASA's Scientist-Astronauts*, 30.
31. Michael Collins, *Carrying the Fire: An Astronaut's Journeys* (New York: Farrar, 1974), 45; Shayler, *NASA's Scientist-Astronauts*, 35.
32. David Sington, "In the Shadow of the Moon" (Discovery Films, 2007).
33. Buzz Aldrin, "Remarks Accompanying Screening of in The Shadow of the Moon" (Washington, DC: Heritage Foundation, September 7, 2007).
34. Slayton and Cassutt, *Deke!*, 143.
35. Cunningham, *The All-American Boys (Revised Edition)*, 36.
36. Eugene Cernan and Don Davis, *The Last Man on the Moon: Astronaut Eugene Cernan and America's Race in Space* (New York: St. Martin's Press, 1999), 187.
37. Atkinson and Shafritz, *The Real Stuff*, 73.

38. National Aeronautics and Space Administration, "NASA Will Recruit 10 to 20 Scientist-Astronauts," *Manned Spacecraft Center Press Release*, October 19, 1964.
39. Shayler, *NASA's Scientist-Astronauts*, 32, 37.
40. Atkinson and Shafritz, *The Real Stuff*, 80, 81.
41. Shayler, *NASA's Scientist-Astronauts*, 55.
42. Courtney G. Brooks, James M. Grimwood, and Loyd S. Swenson, *Chariots for Apollo: A History of Manned Lunar Spacecraft* (Washington, DC: NASA, 1979), 180.
43. Edward G. Gibson, "Oral History Transcript (Carol Butler, Interviewer)," NASA Johnson Space Center Oral History Project (Houston, Texas, December 1, 2000), 4, 11.
44. Shayler, *NASA's Scientist-Astronauts*, 56.
45. Gibson, "Oral History Transcript (Carol Butler, Interviewer)," 7.
46. David J. Shayler, *Skylab: America's Space Station* (Chichester, UK: Praxis, 2001), 106–107.
47. Atkinson and Shafritz, *The Real Stuff*, 81.
48. Gibson, "Oral History Transcript (Carol Butler, Interviewer)," 8, 12, 17–18.
49. Ibid., 28.
50. Shayler, *NASA's Scientist-Astronauts*, 93–97.
51. Cunningham, *The All-American Boys (Revised Edition)*, 291.
52. "...[M]ost common sailors were rather contemptuous of the scientific 'idlers,' as they called anyone who did not stand watch." Rozwadowski, "Small World: Forging a Scientific Maritime Culture for Oceanography," *Isis* 87 (1996): 409–429.
53. Joseph P. Kerwin, "Oral History Transcript (Kevin M. Rusnak, Interviewer)," NASA Johnson Space Center Oral History Project (Houston, Texas, May 12, 2000), 5.
54. Gibson, "Oral History Transcript (Carol Butler, Interviewer)," 12.
55. Atkinson and Shafritz, *The Real Stuff*, 185.
56. Shayler, *NASA's Scientist-Astronauts*, 253.
57. Brian O'Leary, *The Making of an Ex-Astronaut* (Boston: Houghton Mifflin, 1970), 80.
58. "Crew Nomenclature," March 14, 1967, Box 068–12, Apollo Series, Johnson Space Center History Collection, University of Houston–Clear Lake; Brooks, Grimwood, and Swenson, *Chariots for Apollo*, 261 at †; Robert L. Crippen, "Oral History Transcript (Rebecca Wright, Interviewer)," NASA Johnson Space Center Oral History Project (Houston, Texas, May 26, 2006), 24–25.
59. Shayler, *NASA's Scientist-Astronauts*, 253.
60. Francis French and Colin Burgess, *In the Shadow of the Moon: A Challenging Journey to Tranquility, 1965–69* (Lincoln: University of Nebraska Press, 2007), 199.
61. Cunningham, *The All-American Boys (Revised Edition)*, 285.
62. Ibid., 112.
63. Ibid., 285.
64. Collins, *Carrying the Fire*, 19–20, 60.
65. Cunningham, *The All-American Boys (Revised Edition)*, 42.
66. M. Scott Carpenter and Kris Stoever, *For Spacious Skies: The Uncommon Journey of a Mercury Astronaut* (Orlando: Harcourt, 2002), 236.
67. Shayler, *Skylab*, 108.
68. Cunningham, *The All-American Boys (Revised Edition)*, 285.

69. "National Aeronautics and Space Act of 1958," in *Exploring the Unknown: Selected Documents in the History of the U.S. Civil Space Program*, ed. John M. Logsdon (Washington, DC: NASA, 1995).
70. Cunningham, *The All-American Boys (Revised Edition)*, 284–285.
71. Maura Phillips Mackowski, *Testing the Limits: Aviation Medicine and the Origins of Manned Space Flight* (College Station: Texas A&M University Press, 2006), 212.
72. Cunningham, *The All-American Boys (Revised Edition)*, 284–285.
73. Popular culture reinforced this stereotype. A space program that would blast "mild-mannered assistant professor Myron Schwartz" to the Moon, was, to author Lois Philmus, "lunar lunacy." Her 1966 mock-history *A Funny Thing Happened on the Way to the Moon* poked fun at bumbling pilots "Sky" Sawyer and "Wrong-Way" Conners, but saved the greatest sarcasm for the academic—Myron—mistakenly chosen by NASA's "Scientist-Astronaut Program" to "be a passenger on America's first moon shot." Lois C. Philmus, *A Funny Thing Happened on the Way to the Moon* (New York: Spartan Books, 1966).
74. Joseph P. Kerwin, "Oral History Transcript (Kevin M. Rusnak, Interviewer)," NASA Johnson Space Center Oral History Project (Houston, Texas, May 12, 2000), 6.
75. Ibid.
76. Shayler, *NASA's Scientist-Astronauts*, 94, 244.
77. Slayton and Cassutt, *Deke!*, 211–212.
78. Shayler, *NASA's Scientist-Astronauts*, 117.
79. Ibid., 118–120, 244.
80. French and Burgess, *In the Shadow of the Moon*, 148–149, 203.
81. Gideon Kunda, *Engineering Culture: Control and Commitment in a High-Tech Corporation* (Philadelphia: Temple University Press, 1992); Robert Zussman, *Mechanics of the Middle Class: Work and Politics among American Engineers* (Berkeley: University of California Press, 1985).
82. Joseph P. Allen, "Oral History Transcript (Jennifer Ross-Nazzal, Interviewer)," NASA Johnson Space Center Oral History Project (Houston: National Aeronautics and Space Administration, January 28, 2003), 2–5.
83. Atkinson and Shafritz, *The Real Stuff*, 80.
84. Allen, "Oral History Transcript (Jennifer Ross-Nazzal, Interviewer)," 10.
85. Patricia A. Santy, *Choosing the Right Stuff: The Psychological Selection of Astronauts and Cosmonauts* (Westport: Praeger, 1994), 34–35.
86. Allen, "Oral History Transcript (Jennifer Ross-Nazzal, Interviewer)," 3, 6.
87. O'Leary, *The Making of an Ex-Astronaut*.
88. Allen, "Oral History Transcript (Jennifer Ross-Nazzal, Interviewer)," 5.
89. "Aside from the extreme discomfort of storms, landlubber scientists found the most unsettling physical aspect of daily lie at sea to be the rigid and hierarchical arrangement of space in ships." Rozwadowski, "Small World: Forging a Scientific Maritime Culture for Oceanography," 414.
90. Wolfe, *The Right Stuff*, 143, 150.
91. Cunningham, *The All-American Boys (Revised Edition)*, 99.
92. Ibid.
93. Allen, "Oral History Transcript (Jennifer Ross-Nazzal, Interviewer)," 8.
94. Shayler, *NASA's Scientist-Astronauts*, 251.
95. Allen, "Oral History Transcript (Jennifer Ross-Nazzal, Interviewer)," 9.
96. Ibid., 10.

97. Ibid., 11.
98. Shayler, *NASA's Scientist-Astronauts*, 96.
99. Ibid., 182.
100. O'Leary, *The Making of an Ex-Astronaut*, 191–192.
101. Shayler, *NASA's Scientist-Astronauts*, 182.
102. Allen, "Oral History Transcript (Jennifer Ross-Nazzal, Interviewer)," 14, 24.
103. Lapp, "Send Computers, Not Men, into Deep Space," 33.
104. Shayler, *Skylab*, 113.
105. Cunningham, *The All-American Boys (Revised Edition)*, 300–301.
106. O'Leary, *The Making of an Ex-Astronaut*, 80; Shayler, *NASA's Scientist-Astronauts*, 247.
107. Shayler, *NASA's Scientist-Astronauts*, 203.
108. O'Leary, *The Making of an Ex-Astronaut*; Shayler, *NASA's Scientist-Astronauts*, 118.
109. O'Leary, *The Making of an Ex-Astronaut*, 118, 165.
110. See, for example, Joseph Ben-David, *The Scientist's Role in Society: A Comparative Study* (Englewood Cliffs: Prentice-Hall, 1971).
111. See, for example, Robert Kohler, *Lords of the Fly: Drosophila Genetics and the Experimental Life* (Chicago: University of Chicago Press, 1994).
112. "Conflict between the goals of scientists and mariners of all levels was deeply rooted in the political culture of ships. Scientists who failed to understand and negotiate the social and political dynamics on board compromised their scientific work." Rozwadowski, "Small World: Forging a Scientific Maritime Culture for Oceanography," 418.
113. O'Leary, *The Making of an Ex-Astronaut*, 198, 200.
114. Wolfe, *The Right Stuff*, 143.
115. O'Leary, "Topics: Science or Stunts on the Moon?"
116. O'Leary, *The Making of an Ex-Astronaut*, 84, 68.
117. Ibid., 220. "Wallich argued constantly with Capt. Leopold McClintock, complaining that they were not employing sounding devices that retrieved bottom samples frequently enough." Rozwadowski, "Small World: Forging a Scientific Maritime Culture for Oceanography," 418.
118. O'Leary, *The Making of an Ex-Astronaut*, 170.
119. Shayler, *NASA's Scientist-Astronauts*, 157, 177, 184–185.
120. Slayton and Cassutt, *Deke!*, 212.
121. Ibid., 211.
122. Shayler, *NASA's Scientist-Astronauts*, 244–245.
123. Ibid., 205.
124. Ibid., 190, 244–246.
125. "Astronaut Resigns to Pursue Science," *New York Times*, August 6, 1969.
126. O'Leary, *The Making of an Ex-Astronaut*, 223.
127. Shayler, *NASA's Scientist-Astronauts*, 247; Cunningham, *The All-American Boys (Revised Edition)*, 290.
128. Arthur Hill, "Scientist-Astronauts Facing Uncertain Future," *Houston Chronicle*, May 23, 1971.
129. Donald Holmquest, "Letter to James C. Fletcher," July 20, 1973, Folder 2, Box 57, George M. Low Papers, Archives and Special Collections, Rensselaer Polytechnic Institute, Troy, New York.
130. Cunningham, *The All-American Boys (Revised Edition)*, 301.
131. Allen, "Oral History Transcript (Jennifer Ross-Nazzal, Interviewer)," 14.

132. Ibid., 18–19.
133. Ibid., 193–194
134. Ibid., 214–215.
135. Shayler, *Skylab*, 123; Slayton and Cassutt, *Deke!*, 271; Richard Witkin, "Scientist Expected to Be Picked for Moon Trip: Space Agency Reported Set to Name a Geologist Move Is Viewed as Attempt to Answer Criticism," *New York Times*, December 12, 1969.
136. "Letter from Homer E. Newell to Morris S. Petersen," May 12, 1971, Folder 004155, NASA Historical Reference Collection, NASA Headquarters, Washington, DC.
137. Shayler, *Skylab*, 123.
138. Slayton and Cassutt, *Deke!*, 271.
139. Cernan and Davis, *The Last Man on the Moon*, 253, 274.
140. Slayton and Cassutt, *Deke!*, 271.
141. Cunningham, *The All-American Boys (Revised Edition)*, 108.
142. Slayton and Cassutt, *Deke!*, 272.
143. Shayler, *NASA's Scientist-Astronauts*, 194, 215.
144. Robert Farquhar, *Fifty Years on the Space Frontier: Halo Orbits, Comets, Asteroids, and More* (Parker: Outskirts Press, 2011)," 33–34.
145. For example, W. David Compton and Charles D. Benson, *Living and Working in Space: A History of Skylab*, (Washington, DC: NASA, 1983).
146. Compton and Benson, *Living and Working in Space*, 219.
147. Shayler, *NASA's Scientist-Astronauts*, 254–255.
148. Slayton and Cassutt, *Deke!*, 252.
149. Joseph P. Allen, "Oral History Transcript #3 (Jennifer Ross-Nazzal, Interviewer)," NASA Johnson Space Center Oral History Project (Washington, DC, March 18, 2004), 1.
150. Compton and Benson, *Living and Working in Space*, 220.
151. "Letter from Homer E. Newell to Morris S. Petersen," May 12, 1971, Folder 004155, NASA Historical Reference Collection, NASA Headquarters, Washington, DC.
152. Shayler, *NASA's Scientist-Astronauts*, 101.
153. Compton and Benson, *Living and Working in Space*, 220.
154. Crippen, "Oral History Transcript (Rebecca Wright, Interviewer)," 8; Karol J. Bobko, "Oral History Transcript (Summer Chick Bergen, Interviewer)," NASA Johnson Space Center Oral History Project (Houston, Texas, February 12, 2002), 8.
155. Shayler, *NASA's Scientist-Astronauts*, 240.
156. Allen, "Oral History Transcript." 6–7.
157. O'Leary, *The Making of an Ex-Astronaut.*
158. Donald H. Peterson, "Oral History Transcript (Jennifer Ross-Nazzal, Interviewer)," NASA Johnson Space Center Oral History Project (Houston, Texas, November 14, 2002), 77–78.
159. Gibson, "Oral History Transcript (Carol Butler, Interviewer)," 89.
160. Shayler, *NASA's Scientist-Astronauts*, 257.
161. O'Leary, *The Making of an Ex-Astronaut*, 148.
162. Ibid., 147–148.
163. Ibid., 148.
164. Allen, "Oral History Transcript #2 (Jennifer Ross-Nazzal, Interviewer)," NASA Johnson Space Center Oral History Project (Washington, DC, March 16, 2004), 2, 6–7.

4 The Man in the Gray Flannel Spacesuit

1. David Sington, "In the Shadow of the Moon" (Discovery Films, 2007).
2. See, generally, Stephen B. Johnson, *The Secret of Apollo: Systems Management in American and European Space Programs* (Baltimore: Johns Hopkins University Press, 2002).
3. Francis French and Colin Burgess, *In the Shadow of the Moon: A Challenging Journey to Tranquility, 1965–1969* (Lincoln: University of Nebraska Press, 2007), 50–52.
4. Ibid., 227.
5. Francis French and Colin Burgess, *Into That Silent Sea: Trailblazers of the Space Era, 1961–1965* (Lincoln: University of Nebraska Press, 2007), 236.
6. Announced to the public in 1963, this parallel air force human spaceflight program would have employed variants of the launch vehicles and spacecraft of NASA's Project Gemini. Occupied by non-NASA military crews, MOL would have been used for reconnaissance, satellite inspection, and other missions.
7. Donald H. Peterson, "Oral History Transcript (Jennifer Ross-Nazzal, Interviewer)," NASA Johnson Space Center Oral History Project (Houston, Texas, November 14, 2002), 11–12.
8. Richard H. Truly, "Oral History Transcript (Rebecca Wright, Interviewer)," NASA Oral History Project (Golden, Colorado, June 16, 2003), 2.
9. Henry W. Hartsfield Jr., "Oral History Transcript (Carol Butler, Interviewer)," NASA Johnson Space Center Oral History Project (Houston, Texas, June 12, 2001), 2.
10. Peterson, "Oral History Transcript (Jennifer Ross-Nazzal, Interviewer)," 4.
11. C. Gordon Fullerton, "Oral History Transcript (Rebecca Wright, Interviewer)," NASA Johnson Space Center Oral History Project (NASA Dryden Flight Research Center, California, May 6, 2002), 6–8. Fullerton's experience in flying large aircraft would later serve him well, but he had been received warily in the fighter-driven test pilot community.
12. Hartsfield, "Oral History Transcript (Carol Butler, Interviewer), 6; Fullerton, "Oral History Transcript (Rebecca Wright, Interviewer)," 8.
13. Truly, "Oral History Transcript (Rebecca Wright, Interviewer)," 2–7, 9.
14. Robert L. Crippen, "Oral History Transcript (Rebecca Wright, Interviewer)," NASA Johnson Space Center Oral History Project (Houston, Texas, May 26, 2006), 3.
15. Hartsfield, "Oral History Transcript (Carol Butler, Interviewer)," 8.
16. Truly, "Oral History Transcript (Rebecca Wright, Interviewer)," 16.
17. Hartsfield, "Oral History Transcript (Carol Butler, Interviewer)," 8. Fullerton, like other astronauts of the era, was simultaneously relieved to have avoided service in Vietnam and guilty over his good fortune. Fullerton, "Oral History Transcript (Rebecca Wright, Interviewer)," 6.
18. Hartsfield, "Oral History Transcript (Carol Butler, Interviewer)," 9.
19. Peterson, "Oral History Transcript (Jennifer Ross-Nazzal, Interviewer)," 12.
20. Donald K. Slayton and Michael Cassutt, *Deke! U.S. Manned Space: From Mercury to the Shuttle* (New York: St. Martin's Press, 1994), 250–251; See also, generally, David J. Shayler, *Skylab: America's Space Station* (Chichester, UK: Praxis, 2001).
21. Hartsfield, "Oral History Transcript (Carol Butler, Interviewer)," 9.
22. Many of the MOL pilots who stayed behind enjoyed long military careers as program managers, undamaged by their association with the never-flown MOL.

Peterson, "Oral History Transcript (Jennifer Ross-Nazzal, Interviewer)," 12–14.

23. Hartsfield, "Oral History Transcript (Carol Butler, Interviewer)," NASA Johnson Space Center Oral History Project (Houston, Texas, June 12, 2001), 21.
24. Fullerton, "Oral History Transcript (Rebecca Wright, Interviewer)," 10.
25. Walter Cunningham, *The All-American Boys (Revised Edition)* (New York: ibooks, 2003), 294.
26. Fullerton, "Oral History Transcript (Rebecca Wright, Interviewer)," 11–12.
27. Howard Muson, "Comedown from the Moon: What Has Happened to the Astronauts," *New York Times Magazine*, December 3, 1972, 139.
28. Slayton and Cassutt, *Deke!*, 266.
29. Cunningham, *The All-American Boys (Revised Edition)*, 262–263.
30. Slayton and Cassutt, *Deke!*, 237.
31. Al Worden, *Falling to Earth: An Apollo Astronaut's Journey* (Washington, DC: Smithsonian Books, 2011), 245–248.
32. Ray E. Wood, "Interview with Captain John W. Young," 1972, Folder 008943, NASA Historical Reference Collection, NASA Headquarters, Washington, DC.
33. "Interrogatory of Dr. Donald L. Holmquest," August 30, 1972, Folder 008943, NASA Historical Reference Collection, NASA Headquarters, Washington, DC.
34. George M. Low, "Handwritten Annotation," September 15, 1972 (typed September 18, 1972) in "Reply to Cover Note from James C. Fletcher to George M. Low," September 14, 1972, attaching article from Jay Russell and William Cromie, "Astronauts Strip All Moon Craft" Tomball Tribune, August 30, 1972, Folder 4, Box 50, George M. Low Papers, Archives and Special Collections, Rensselaer Polytechnic Institute, Troy, New York.
35. Donald Holmquest, "Letter to James C. Fletcher, Eisenhower Medical Center," July 20, 1973. George M. Low Papers. Archives and Special Collections, Rensselaer Polytechnic Institute, Troy, N.Y., Box 57, Folder 2, 3.
36. Muson, "Comedown from the Moon," 139.
37. Holmquest, "Letter to James C. Fletcher," 2–3.
38. For example, Jerry Bledsoe, "Down from Glory," *Esquire*, January 1973.
39. W. David Compton and Charles D. Benson, *Living and Working in Space: A History of Skylab* (Washington, DC: NASA, 1983), 221.
40. David J. Shayler, *NASA's Scientist-Astronauts* (New York: Springer, 2007), 254.
41. Karol J. Bobko, "Oral History Transcript (Summer Chick Bergen, Interviewer)," NASA Johnson Space Center Oral History Project (Houston, Texas, February 12, 2002), 9; George M. Low, "Handwritten Note to ADA Shapley," September 25, 1972 (typed September 26, 1972), affixed to "Memo to George M. Low from Dale D. Myers Re: Skylab Medical Experiments Altitude Test Completion," September 22, 1972, Folder 4, Box 50, George M. Low Papers, Archives and Special Collections, Rensselaer Polytechnic Institute, Troy, New York.
42. Truly, "Oral History Transcript (Rebecca Wright, Interviewer)," 12–13.
43. Hartsfield, "Oral History Transcript (Carol Butler, Interviewer)," 34.
44. Compton and Benson, *Living and Working in Space*, 289, 310.
45. George M. Low, "Memorandum to James Fletcher Re: Skylab Review in Houston," August 2 1973, Folder 004157, NASA Historical Reference Collection, NASA Headquarters, Washington, DC.

46. Gibson, "Oral History Transcript (Carol Butler, Interviewer)," 58.
47. Ibid., 59–60.
48. Ibid., 49, 61, 57, 68.
49. Compton and Benson, *Living and Working in Space*, 280–281.
50. Ibid., 60–62.
51. Ibid., 65–66.
52. Ibid., 84, Molly Ivins, "Ed Who?," *New York Times*, June 30, 1974, 14.
53. Shayler, *Skylab*, 304–305; David Shayler, *Apollo: The Lost and Forgotten Missions*, Springer-Praxis Books in Astronomy and Space Sciences (Chichester: Springer, 2002). In this exercise in counterfactual history, amateur historians publishing on the Internet have done some impressive work, positing possible prime, backup, and support crews for each mission through the hypothetical Skylab 9 mission, based on the pool of available astronauts and the assignment conventions of the Astronaut Office. Mark Wade, "Your Flight Has Been Cancelled," Encyclopedia Astronautica, http://www.astronautix.com/articles/youelled.htm (accessed April 4, 2010).
54. Thomas O. Paine, "Letter to Robert R. Gilruth," February 16, 1970, Folder 008948, NASA Historical Reference Collection, NASA Headquarters, Washington, DC.
55. Fletcher described to Low how he respected Conrad's effort to communicate his concerns, but Fletcher feared, as Kraft had warned, that future discussions with Conrad would be "vitriolic" and "confused." James C. Fletcher, "Memorandum to George E. Low Re: Resignation of Pete Conrad," October 17, 1973, Folder 3, Box 57, George M. Low Papers, Archives and Special Collections, Rensselaer Polytechnic Institute, Troy, New York.
56. Shayler, *NASA's Scientist-Astronauts*, 196, 241, 287.
57. By the mid-1970s, Skylab's orbit was already decaying, and, without boosting, the station would likely reenter the atmosphere, potentially striking a populated area. Rookie pilot-astronauts Vance Brand and Don Lind trained to dock with the empty Skylab and fire the Propulsion System of their Apollo Service Module to lower Skylab's orbital altitude. After firing the engine, Brand and Lind would have had only 14 minutes to close the hatch, undock, and withdraw from Skylab before being dragged down with it or struck by debris as Skylab began to break up in the upper atmosphere. The STS-2A "Skylab Reboost," an early space shuttle flight scheduled for 1979, would have sent veteran astronauts Fred Haise and Jack Lousma back to Skylab to boost the station into a higher orbit. Unfortunately for the astronauts, repeated delays pushed the first shuttle flight from 1978 to 1981. By then, Skylab had reentered Earth's atmosphere and Haise had retired. Ultimately, neither mission was flown. Shayler, *Skylab*, 298–299, 308.
58. Shayler, *NASA's Scientist-Astronauts*, 279.
59. Slayton and Cassutt, *Deke!*, 277.
60. Ibid., 280–281.
61. Cunningham, *The All-American Boys (Revised Edition)*, 330.
62. Bobko, "Oral History Transcript (Summer Chick Bergen, Interviewer)," 11.
63. Cunningham, *The All-American Boys (Revised Edition)*, 340–343; Slayton and Cassutt, *Deke!*, 304–305; Thomas P. Stafford and Michael Cassutt, *We Have Capture: Tom Stafford and the Space Race* (Washington, DC: Smithsonian Institution Press, 2002), 193–194.
64. Ibid.

65. Cunningham and Herskowitz, *The All-American Boys*, 290.
66. Ibid., 292. 25 years later, though, an older, wiser Cunningham placed his comments in the past tense. By the time Cunningham published a revised and updated version of his memoir in 2003, his opinion on older astronauts had softened considerably. He concluded a reprise of his earlier comments with a short commentary on the attitudes he had earlier expressed, and noted John Glenn's shuttle flight at the age of 77: "That representing our thinking in the Seventies!" Cunningham, *The All-American Boys (Revised Edition)*, 340.
67. Chester M. Lee, "Letter to George Abbey, et al.," August 16, 1978, Folder 008958, NASA Historical Reference Collection, NASA Headquarters, Washington, DC.
68. James A. Loudon, "Why We Really Want a Space Shuttle," *New York Times*, March 28, 1972, 42.
69. "How on Earth did you ever—why would you ever think of this?" veteran astronaut T. K. Mattingly is said to have responded. Donald H. Peterson, "Oral History Transcript (Jennifer Ross-Nazzal, Interviewer)," NASA Johnson Space Center Oral History Project (Houston: National Aeronautics and Space Administration, November 14, 2002),36.
70. For Haise, who, following Apollo 13's return, had been badly burned (and according to Allen, nearly killed) in an airplane crash, the chance to fly *Enterprise* was particularly fortunate. Joseph P. Allen, "Oral History Transcript #2 (Jennifer Ross-Nazzal, Interviewer)," NASA Johnson Space Center Oral History Project (Washington, DC: National Aeronautics and Space Administration, March 16, 2004), 18.
71. Fullerton, "Oral History Transcript (Rebecca Wright, Interviewer)," 15.
72. Peterson, "Oral History Transcript (Jennifer Ross-Nazzal, Interviewer)," 56.
73. Bobko, "Oral History Transcript (Summer Chick Bergen, Interviewer)," 15.
74. "Science: Commuting in Space," *Time*, December 15, 1975.
75. Hartsfield, "Oral History Transcript (Carol Butler, Interviewer)," 27–29.
76. Fullerton, "Oral History Transcript (Rebecca Wright, Interviewer)," 19, 21.
77. Peterson, "Oral History Transcript (Jennifer Ross-Nazzal, Interviewer), "46–49.
78. Ibid., 54, Martin Campbell-Kelly and William Aspray, *Computer: A History of the Information Machine* (Boulder: Westview Press, 2004).
79. Peterson, "Oral History Transcript (Jennifer Ross-Nazzal, Interviewer)," 55.
80. Fullerton, "Oral History Transcript (Rebecca Wright, Interviewer)," 21–22.
81. Christopher C. Kraft, "Letter to Elmer S. Groo," April 15, 1977, Folder 008948, NASA Historical Reference Collection, NASA Headquarters, Washington, DC.
82. Peterson, "Oral History Transcript (Jennifer Ross-Nazzal, Interviewer)."
83. Cunningham, *The All-American Boys (Revised Edition)*, 350.
84. During the mid-1970s Young and fellow Apollo veteran T. K. Mattingly forwarded regular memos to NASA managers about unremedied deficiencies in the space shuttle that might gravely impact crew safety. See, for example, "Space Shuttle Costs and Schedules #2," October 10, 1974, Box 1, Thomas K. Mattingly Files, Center Series, Johnson Space Center History Collection, University of Houston–Clear Lake.
85. Ibid., 352.
86. Ibid.
87. Ibid., 348–349, 351, 361.

88. Mattingly, "Oral History Transcript #2 (Kevin M. Rusnak, Interviewer)," 2–3.
89. Andrew Chaikin, "George Abbey: NASA's Most Controversial Figure," *Space.com*, February 26, 2001.
90. Cunningham, *The All-American Boys (Revised Edition)*, 360–362.
91. Chaikin, "George Abbey: NASA's Most Controversial Figure."
92. Hartsfield, "Oral History Transcript (Carol Butler, Interviewer)," 30.
93. R. Mike Mullane, *Riding Rockets: The Outrageous Tales of a Space Shuttle Astronaut* (New York: Scribner, 2006), 25, Carl Walz, "Astronaut Adventures" (paper presented at the Smithsonian Institution Folklife Festival, Washington, DC, June 26 2008).
94. Crippen, "Oral History Transcript (Rebecca Wright, Interviewer)," 1. Fullerton recalled that Abbey had offered him a position on the shuttle Enterprise's landing and approach tests in a similar manner. Fullerton, "Oral History Transcript (Rebecca Wright, Interviewer)," 14.
95. Crippen, "Oral History Transcript (Rebecca Wright, Interviewer)," 2.
96. Hartsfield, "Oral History Transcript (Carol Butler, Interviewer)," 30.
97. Hartsfield, "Oral History Transcript #2 (Carol Butler, Interviewer)," NASA Johnson Space Center Oral History Project (Houston, Texas, June 15, 2001), 4.
98. Fullerton, "Oral History Transcript (Rebecca Wright, Interviewer)," 30.
99. Allen M. Steele, *Orbital Decay* (New York: Ace Books, 1989).
100. Charles A. Reich, *The Greening of America* (New York: Random House, 1970), 89, et seq.
101. Patricia A. Santy, *Choosing the Right Stuff: The Psychological Selection of Astronauts and Cosmonauts* (Westport: Praeger, 1994), 22.
102. Joseph D. Atkinson and Jay M. Shafritz, *The Real Stuff: A History of NASA's Astronaut Recruitment Program* (New York: Praeger, 1985), 30.
103. Sheldon J. Korchin and George E. Ruff, "Personality Characteristics of the Mercury Astronauts," in *The Threat of Impending Disaster, Contributions to the Psychology of Stress,* ed. George H. Grosser, Henry Wechsler, and Milton Greenblatt (Cambridge: MIT Press, 1964), 200.
104. Ibid.
105. French and Burgess, *In the Shadow of the Moon*, 136.
106. Thomas Kenneth Mattingly, II, "Oral History Transcript #2 (Kevin M. Rusnak, Interviewer)," NASA Johnson Space Center Oral History Project (Houston, Texas, April 22, 2002), 5–6.
107. French and Burgess, *In the Shadow of the Moon*, 229.
108. Cunningham, *The All-American Boys (Revised Edition)*, 310–311.
109. Michael Collins, *Carrying the Fire: An Astronaut's Journeys* (New York: Farrar, 1974), 462.
110. Ibid., 460.
111. Muson, "Comedown from the Moon," 133.
112. Cunningham, *The All-American Boys (Revised Edition)*, 315.
113. Muson, "Comedown from the Moon," 37.
114. Collins, *Carrying the Fire*, 461.
115. See, generally, Matthew H. Hersch, "Space Madness: The Dreaded Disease that Never Was," Endeavour 36 (2012): 32–40.
116. Edwin E. Aldrin and Wayne Warga, *Return to Earth* (New York: Random House, 1973), 295, et seq.; Cunningham, *The All-American Boys (Revised Edition)*, 168.

117. Aldrin and Warga, *Return to Earth*, 298–300.
118. Ibid., 297; Collins, *Carrying the Fire*, 460–461.
119. Muson, "Comedown from the Moon," 134.
120. Cunningham, *The All-American Boys (Revised Edition)*, 168.
121. Worden, *Falling to Earth*, 243–244.
122. Muson, "Comedown from the Moon," 140.
123. Frank Borman, Jim Lovell, and Bill Anders, "John H. Glenn Lecture: An Evening with the Apollo 8 Astronauts," Smithsonian Institution, National Air and Space Museum, November 13, 2008.
124. For example, Bledsoe, "Down from Glory."
125. Muson, "Comedown from the Moon," 132.
126. Ibid., 37.
127. William Friedkin, "The Exorcist" (Hoya Productions, 1973).
128. William Peter Blatty, "The Ninth Configuration" (Warner Bros., 1980).
129. James B. Irwin, *More Than Earthlings: An Astronaut's Thoughts for Christ-Centered Living* (Nashville: Broadman Press, 1983).
130. Ray Bradbury, "The Rocket Man," in *The Illustrated Man* (Garden City: Doubleday & Company, 1951).
131. Byron Haskin, "Conquest of Space" (Paramount Pictures Corp., 1955).
132. Sloan Wilson, *The Man in the Gray Flannel Suit* (New York: Simon and Schuster, 1955).
133. Nicholas de Monchaux, *Spacesuit: Fashioning Apollo* (Cambridge, Massachusetts: MIT Press, 2011), 51 (citing "The Bird and the Watcher," *Time*, April 1, 1957: 8).
134. This particular joke has been decades in the making. See, for example, Thomas Mallon, "Satellite of Love," *New York Times*, April 9, 2006, F7.

5 Public Space

1. "Trekkies (or Trekkers)," http://www.tvacres.com/fans_trekkies.htm (accessed April 4, 2010); "Timeline," *USS Stargazer II Domain*, http://www.stargazertwo.com/Database/DS9/Timeline.htm (accessed April 4, 2010).
2. Ed Papazian, *Medium Rare: The Evolution, Workings, and Impact of Commercial Television* (New York: Media Dynamics, 1991), 241–244.
3. William Grimes, "Joan Winston, 'Trek' Superfan, Dies at 77," *New York Times*, September 21, 2008, 34; Joan Winston, *The Making of the Trek Conventions: Or, How to Throw a Party for 12,000 of Your Most Intimate Friends* (New York: Doubleday, 1977), 26.
4. Francis French and Colin Burgess, *Into That Silent Sea: Trailblazers of the Space Era, 1961–1965* (Lincoln: University of Nebraska Press, 2007), 266 [quoting authors' interview with Cooper].
5. NASA Space Task Group, "Handwritten Revisions to 12/30/1960 Telex from Loudon Wainwright, Life Magazine," Records of the National Aeronautics and Space Administration. National Archives and Records Center at College Park, Maryland.
6. Michael Collins, *Carrying the Fire: An Astronaut's Journeys* (New York: Farrar, 1974), 76.
7. Francis French and Colin Burgess, *In the Shadow of the Moon: A Challenging Journey to Tranquility, 1965–1969* (Lincoln: University of Nebraska Press, 2007), 7, 44.

8. For example, Peter Wolfe, *In the Zone: The Twilight World of Rod Serling* (Bowling Green: Bowling Green State University Popular Press, 1997).
9. Don Medford, "Death Ship," *Twilight Zone* (CBS, February 7, 1963).
10. For example, Wolfe, *In the Zone*, 100, 125.
11. Lewis Gilbert, "You Only Live Twice" (United Artists, 1967).
12. Lewis Gilbert, "Moonraker" (United Artists, 1979).
13. George E. Low, "Letter to Edward E. David," October 30, 1970, Folder 004154, NASA Historical Reference Collection, NASA Headquarters, Washington, DC.
14. "...NASA was created expressly for the purpose of conducting peaceful space missions, and the orbiting of a military astronaut will be identified by the world in general as a military gesture, and is sure to be seized upon by the U.S.S.R. for propaganda purposes." Jerome B. Weisner, "Memorandum for Dr. Bundy," March 9, 1961, Box 3, Flight Crew Operations Directorate, Center Series, Johnson Space Center History Collection, University of Houston–Clear Lake.
15. Stephen Cox, *Dreaming of Jeannie: TV's Prime Time in a Bottle* (New York: St. Martin's Griffin, 2000), 58.
16. Shepard, though, vigorously denied any similarity. Ibid., 60–62.
17. Ibid., 58–61.
18. Tom Wolfe, *The Right Stuff* (New York: Farrar, Straus, and Giroux, 1979), 120.
19. On the gendering of military activities, see, Joan Wallach Scott, "Gender: A Useful Category of Historical Analysis," in *Gender and the Politics of History* (New York: Columbia University Press, 1999), 48. See, also, Carol Cohn, "Wars, Wimps, and Women: Talking Gender and Thinking War," in *Gendering War Talk*, ed. Miriam Cooke and Angela Woollacott (Princeton: Princeton University Press, 1993).
20. Howard E. McCurdy, *Space and the American Imagination* (Washington, DC: Smithsonian Institution Press, 1997), 91.
21. As Vivian Sobchack writes, astronauts in the media exuded a "virginal ideal"—sober, stolid, and sexless—that appealed to audiences who found them both inspiring and safe. Vivian Sobchack, "The Virginity of Astronauts: Sex and the Science Fiction Film," in *Alien Zone: Cultural Theory and Contemporary Science Fiction Cinema*, ed. Annette Kuhn (New York: Verso, 1990), 108.
22. For example, Roger D. Launius, *NASA: A History of the U.S. Civil Space Program* (Malabar: Krieger Pub. Co., 1994), 94.
23. "The Moon Landing," *New York Times*, July 18, 2009, A20.
24. National Aeronautics and Space Administration, *Apollo 11 Technical Air-to-Ground Voice Transcription (GOSS NET 1)* (Houston: Manned Spacecraft Center, 1969), 609.
25. Ralph E. Lapp, "Send Computers, Not Men, into Deep Space," *New York Times*, February 2, 1969, 38.
26. Peder Anker, "The Ecological Colonization of Space," *Environmental History* 10 (2005): 240–244.
27. R. Buckminster Fuller, *Operating Manual for Spaceship Earth* (Carbondale: Southern Illinois University Press, 1969); Anker, "The Ecological Colonization of Space," 244–245, fn. 221.
28. Sheila Jasanoff, "Image and Imagination," in *Changing the Atmosphere: Expert Knowledge and Environmental Governance*, ed. Clark A. Miller and Paul N. Edwards (Cambridge, Massachusetts: MIT Press, 2001), 318.

29. For example, Mel Horwitch, *Clipped Wings: The American SST Conflict* (Cambridge, Massachusetts: MIT Press, 1982); Erik M. Conway, *High-Speed Dreams: NASA and the Technopolitics of Supersonic Transportation, 1945–1999* (Baltimore: Johns Hopkins University Press, 2005).
30. Collins, *Carrying the Fire*, 317.
31. Lapp, "Send Computers, Not Men, into Deep Space," 35.
32. For example, Jasanoff, "Image and Imagination."
33. Fred Turner, *From Counterculture to Cyberculture: Stewart Brand, the Whole Earth Network, and the Rise of Digital Utopianism* (Chicago: University of Chicago Press, 2006), 69.
34. Michael Hohl, "Steward Brand, 'Whole-Earth' Buttons, 1966: The First Photograph of the Whole Earth from Space," http://hohlwelt.com/en/interact/context/sbrand.html (accessed April 4, 2010).
35. Roberta Hornig and James Welsh, "A World in Danger 2: Pollution Totals Ton a Year for Each of Us," *Washington Evening Star*, January 22, 1970, NASA Historical Reference Collection, NASA Headquarters, Washington, DC.
36. See, for example, Jennifer Levasseur, " 'Here's the Earth Coming up': Analysis of the Apollo 8 'Earthrise' Photograph" (paper presented at the *Annual Meeting of the Society for the History of Technology*, Lisbon, Portugal, October 12, 2008).
37. Portola Institute, *The Last Whole Earth Catalog: Access to Tools* (New York: Random House, 1971).
38. Seth Borenstein, "Astronauts Recall View before Earth Day: Space Travelers Recall What It's Like to See Their Home Planet from above Ahead of Earth Day," *Associated Press*, April 21, 2007.
39. Hornig and Welsh, "A World in Danger 2: Pollution Totals Ton a Year for Each of Us."
40. Popular among radical artists of the period, these devices featured prominently in television coverage of the Apollo Moon landings, where they often looked less like scientific instruments than six-shooters wielded by "space cowboys." Constance M. Lewallen and Steve Seid, *Ant Farm 1968–1978* (Berkeley: University of California Press, 2004), plate 1 ("Space Cowboy Meets Plastic Businessman," 1969, performance at Alley Theater, Houston).
41. French and Burgess, "In the Shadow of the Moon," 311–312.
42. David Sington, "In the Shadow of the Moon" (Discovery Films, 2007).
43. Neil Maher, "Neil Maher on Shooting the Moon," *Environmental History* 9 (2004): 526–531; National Aeronautics and Space Administration, "Apollo 17 30th Anniversary: Antarctica Zoom-out," http://svs.gsfc.nasa.gov/vis/a000000/a002600/a002680/ (accessed April 4, 2010); Jasanoff, "Image and Imagination," 316–317.
44. George M. Low, "Letter to Brian O'Brion," January 10, 1972, p. 1, Folder 004156, NASA Historical Reference Collection, NASA Headquarters, Washington, DC.
45. See, Pamela Etter Mack, *Viewing the Earth: The Social Construction of the Landsat Satellite System* (Cambridge, Massachusetts: MIT Press, 1990).
46. John Donnelly, "Letter to James Fletcher," January 20, 1973, p. 1, Folder 004157, NASA Historical Reference Collection, NASA Headquarters, Washington, DC.
47. Ibid., 4.

48. James C. Fletcher, "Cover Memorandum to George M. Low," February 21, 1973, Folder 004157, NASA Historical Reference Collection, NASA Headquarters, Washington, DC.
49. For example, Richard D. Johnson and Charles Holbrow, eds., *Space Settlements: A Design Study* (Washington DC: NASA, 1977).
50. De Witt Douglas Kilgore, *Astrofuturism: Science, Race, and Visions of Utopia in Space* (Philadelphia: University of Pennsylvania Press, 2003), 161, 165.
51. Douglas Trumbull, "Silent Running" (Universal Pictures, 1972).
52. Andrei Tarkovsky, "Solyaris" (Mosfilm, 1972).
53. Molly Ivins, "Ed Who?," *New York Times*, June 30, 1974, 12.
54. Russell Schweickart, "Space Colonies Should Keep away from the Government for a While," in *CoEvolution Quarterly* 9 (Spring 1976): 72–78; *Space Colonies* (available from http://settlement.arc.nasa.gov/CoEvolutionBook/) (accessed February 20, 2012). See, also, Fred Turner, *From Counterculture to Cyberculture: Stewart Brand, the Whole Earth Network, and the Rise of Digital Utopianism* (Chicago: University of Chicago Press, 2006), 126–127.
55. "Hearing before the Committee on Science and Astronautics," April 23, 1974, Folder 004158, NASA Historical Reference Collection, NASA Headquarters, Washington, DC.
56. Ivins, "Ed Who?,", 14.
57. Mark D. Bowles, *Science in Flux: NASA's Nuclear Program at Plum Brook Station, 1955–2005* (Washington, DC: NASA, 2006), 219.
58. Gerard J. De Groot, *Dark Side of the Moon: The Magnificent Madness of the American Lunar Quest* (New York: New York University Press, 2006), 244–245.
59. Howard Muson, "Comedown from the Moon: What Has Happened to the Astronauts," *New York Times Magazine*, December 3, 1972, 132.
60. Bowles, *Science in Flux*, 224, 227.
61. Kim McQuaid, "Earthly Environmentalism and the Space Exploration Movement, 1960–1990: A Study in Irresolution," *Space Policy* 26 (2010): 163–173.
62. For example, John M. Logsdon, "Project Apollo: Americans to the Moon," in *Exploring the Unknown: Selected Documents in the History of the U.S. Civil Space Program*, ed. John M. Logsdon and Roger D. Launius (Washington, DC: NASA, 2008), 409–419; Edward Clinton Ezell and Linda Neuman Ezell, *The Partnership: A History of the Apollo-Soyuz Test Project* (Washington DC: National Aeronautics and Space Administration, 1978), 50–55.
63. George E. Low, "Letter to Edward E. David," October 30, 1970, Folder 004154, NASA Historical Reference Collection, NASA Headquarters, Washington, DC.
64. David J. Shayler, *Skylab: America's Space Station* (Chichester, UK: Praxis, 2001), 59.
65. Martin Caidin, *Marooned* (New York: Dutton, 1964), John Sturges, "Marooned" (Columbia Pictures Corporation, 1969).
66. Ezell and Ezell, *The Partnership: A History of the Apollo-Soyuz Test Project*, 9–10 (quoting Handler to Paine, 28 May 1970; Handler to Ezell, October 1974; and Handler, "Trip Report").
67. See, for example, Launius, *NASA: A History of the U.S. Civil Space Program*, 107–109; Caspar W. Weinberger, "Memorandum for the President," in *Exploring the Unknown: Selected Documents in the History of the U.S. Civil Space Program*, ed. John M. Logsdon (Washington, DC: NASA, 1995).

68. President Richard M. Nixon, "The Statement by President Nixon, 5 January 1972," National Aeronautics and Space Administration, http://history.nasa.gov/stsnixon.htm (accessed April 4, 2010).
69. Howard E. McCurdy, *Inside NASA: High Technology and Organizational Change in the U.S. Space Program* (Baltimore: Johns Hopkins University Press, 1993), 85–86, 141–146.
70. David J. Shayler, *NASA's Scientist-Astronauts* (New York: Springer, 2007), 285.
71. James A. Loudon, "Why We Really Want a Space Shuttle," *New York Times*, March 28, 1972, 42.
72. Spacelab would be built by the European Space Agency.
73. James C. Fletcher, "Statement by Dr. Fletcher, NASA Administrator," National Aeronautics and Space Administration, http://history.nasa.gov/stsnixon.htm (accessed April 4, 2010).
74. T. A. Heppenheimer, *Development of the Shuttle, 1972–1981* (Washington DC: Smithsonian Institution Press, 2002), 100–101.
75. Brian O'Leary, "Do We Really Want a Space Shuttle?," *New York Times*, February 16, 1972.
76. Brian O'Leary, "Space Shuttle: Billions for a Boondoggle?," *New York Times*, May 16, 1972.
77. O'Leary, "Do We Really Want a Space Shuttle?"
78. O'Leary, "Space Shuttle: Billions for a Boondoggle?"; O'Leary, "Do We Really Want a Space Shuttle?"
79. Joseph D. Atkinson and Jay M. Shafritz, *The Real Stuff: A History of NASA's Astronaut Recruitment Program* (New York: Praeger, 1985), 149–150.
80. Shayler, *NASA's Scientist-Astronauts*, 335.
81. Ibid., 296–297.
82. Allen, "Oral History Transcript (Jennifer Ross-Nazzal, Interviewer)," 9.
83. Shayler, *NASA's Scientist-Astronauts*, 196, 287.
84. Heppenheimer, *Development of the Shuttle, 1972–1981*, 388.
85. George M. Low, "Handwritten Annotations" May 2, 1972 (typed May 4, 1972) on "Memorandum from General Smart to George Low Re: Post Skylab Assignment for Astronauts," George M. Low Papers, Folder 1, Box 51, Archives and Special Collections, Rensselaer Polytechnic Institute, Troy, New York.
86. "Report of the Subcommittee on Scientists Astronauts of the NASA Space Program Advisory Council" (1975), NASA Historical Reference Collection, NASA Headquarters, Washington, DC, 8.
87. Frederick Seitz, "Letter to George M. Low," September 8, 1975, Folder 004159, NASA Historical Reference Collection, NASA Headquarters, Washington, DC.
88. "Report of the Subcommittee on Scientists Astronauts of the NASA Space Program Advisory Council." Folder 004159, NASA Historical Reference Collection, NASA Headquarters, Washington, DC, 4, 7.
89. Atkinson and Shafritz, *The Real Stuff*, 139.
90. Shayler, *NASA's Scientist-Astronauts*, 300.
91. "Report of the Subcommittee on Scientists Astronauts of the NASA Space Program Advisory Council," 13 note. Folder 004159, NASA Historical Reference Collection, NASA Headquarters, Washington, DC.
92. "Meeting Record Re: Shuttle Crew Selection, Fletcher, et al.," January 23, 1975, Folder 1, Box 104, George M. Low Papers, Archives and Special Collections, Rensselaer Polytechnic Institute, Troy, New York.

93. Christopher C. Kraft, "Letter to John E. Naugle," September 3, 1975, Folder 008961, NASA Historical Reference Collection, NASA Headquarters, Washington, DC.
94. Following the arrival of Mission Specialists in the 1978, NASA attempted temporarily to redesignate the remaining 1965 and 1967 scientist-astronauts as "Senior Scientist-Astronauts," but the title did not stick. Shayler, *NASA's Scientist-Astronauts*, 328.
95. "Report of the Subcommittee on Scientists Astronauts of the NASA Space Program Advisory Council." Folder 004159, NASA Historical Reference Collection, NASA Headquarters, Washington, DC, 9–10.
96. Shayler, *NASA's Scientist-Astronauts*, 287.
97. B. R. McCullar, "Meeting Record, Mission Specialists/Payload Specialists Policy (Naugle, Hinners Schweickart, Chapelle, Kennedy, Glaser), 12:30pm," November 9, 1976, Folder 008961. NASA Historical Reference Collection, NASA Headquarters, Washington, DC.
98. Lyn Cywanowicz, "NASA Expands Payload Specialist Opportunities," *NASA News (Marshall Space Flight Center)*, October 22, 1982.
99. Henry W. Hartsfield, Jr., "Oral History Transcript (Carol Butler, Interviewer)," NASA Johnson Space Center Oral History Project (Houston, Texas, June 12, 2001), 40–41.
100. Edward P. Andrews, "Meeting Record Re: Payload Specialist/Mission Specialist and Pilot Astronauts," October 19, 1976, Folder 008961, NASA Historical Reference Collection, NASA Headquarters, Washington, DC.
101. Ibid.
102. John E. Naugle, "Memorandum to James Fletcher Re: Payload Specialists," February 27, 1978, Folder 008961, NASA Historical Reference Collection, NASA Headquarters, Washington, DC.
103. James C. Fletcher, "Letter to Hans M. Mark," November 3, 1976, Folder 008960, NASA Historical Reference Collection, NASA Headquarters, Washington, DC; "Shuttle Crew Selection Change Worries Scientific Community," *Aviation Week & Space Technology*, August 7, 1978.
104. Christopher C. Kraft, "Some JSC Concerns About Payload Specialists," May 10, 1978, Folder 008960, NASA Historical Reference Collection, NASA Headquarters, Washington, DC, 1–2.
105. Christopher C. Kraft, "Letter to Alan M. Lovelace," May 10, 1978, p. 1, Folder 008960, NASA Historical Reference Collection, NASA Headquarters, Washington, DC.
106. Shayler, *NASA's Scientist-Astronauts*, 297–298, 304–305.
107. Thomas O'Toole, "NASA Gives Police Power to Spaceship Captains," *The Washington Post*, March 18, 1980, A8.
108. Amy Elizabeth Foster, *Integrating Women into the Astronaut Corps: Politics and Logistics, 1972–2004* (Baltimore: Johns Hopkins University Press, 2011), 48.
109. Atkinson and Shafritz, *The Real Stuff*, 95–96; Francis French and Colin Burgess, *Into That Silent Sea*, 134; Bettyann Kevles, *Almost Heaven: The Story of Women in Space* (New York: Basic Books, 2003); Margaret A. Weitekamp, *Right Stuff, Wrong Sex: America's First Women in Space Program* (Baltimore: Johns Hopkins University Press, 2004); Foster, *Integrating Women into the Astronaut Corps*, 47–48; Maura P. Mackowski, *Testing the Limits: Aviation Medicine and the Origins of Manned Space Flight* (College Station: Texas A&M University Press, 2006), 206.

110. Collins, *Carrying the Fire*, 178.
111. Mackowski, *Testing the Limits*, 213.
112. Foster, *Integrating Women into the Astronaut Corps*, 51.
113. Kevles, *Almost Heaven: The Story of Women in Space*, 41.
114. French and Burgess, *Into That Silent Sea*, 44.
115. Kevles, *Almost Heaven: The Story of Women in Space*, 16–17.
116. Collins, *Carrying the Fire*, 178.
117. Atkinson and Shafritz, *The Real Stuff*, 90, 103.
118. Kevles, *Almost Heaven: The Story of Women in Space*, 14.
119. While it was the policy of NASA not to disclose the reasons for its selection decisions, Robert Gilruth later explained in an internal memorandum that Dwight's fitness reports and academic credentials were less competitive than those of other applicants. For example, "Memorandum from Robert Gilruth Re: Captain Edward J. Dwight, USAF," April 16, 1965, Box 2, Flight Crew Operations Directorate, Center Series, Johnson Space Center History Collection, University of Houston–Clear Lake.
120. Atkinson and Shafritz, *The Real Stuff*, 102.
121. Ibid., 100.
122. This spoken-word performance was written and recorded by Gil Scott-Heron in 1970.
123. Foster, *Integrating Women into the Astronaut Corps*, 37.
124. Atkinson and Shafritz, *The Real Stuff*, 135; Foster, *Integrating Women into the Astronaut Corps*, 51.
125. "Renaming Of The Manned Spacecraft Center," February 27, 1973, Box 074–31, Apollo Series, Johnson Space Center History Collection, University of Houston–Clear Lake.
126. Atkinson and Shafritz, *The Real Stuff*, 136–137.
127. Ibid., 138.
128. Brian B. King, "Letter to George M. Low," December 4, 1973, Folder 004157, NASA Historical Reference Collection, NASA Headquarters, Washington, DC.
129. George M. Low, "Letter to Brian B. King," December 28, 1973, p. 2, Folder 004157, NASA Historical Reference Collection, NASA Headquarters, Washington, DC.
130. Atkinson and Shafritz, *The Real Stuff*, 139.
131. Robert Reinhold, "Shuttle Mission Puts Focus on Research Crewmen," *New York Times*, November 13, 1982.
132. Atkinson and Shafritz, *The Real Stuff*, 140–141, 148–149.
133. Foster, *Integrating Women into the Astronaut Corps*, 91, 100–102.
134. Atkinson and Shafritz, *The Real Stuff*, 140.
135. Homer Hickam, "What Makes an Astronaut Crack," *Los Angeles Times*, February 9, 2007.
136. "The dynamic shipboard encounter of sailors and scientists resulted in a modified and tamed version of maritime culture." Helen M. Rozwadowski, "Small World: Forging a Scientific Maritime Culture for Oceanography," *Isis* 87 (1996): 409–429.
137. Donald K. Slayton and Michael Cassutt, *Deke! U.S. Manned Space: From Mercury to the Shuttle* (New York: St. Martin's Press, 1994). 314.

Conclusion

1. Dick Siegel, "Rip Van Spaceman: First American Spaceman Returns after 47 Years!," *Weekly World News*, July 25, 2005, 24–25.
2. See, generally, Matthew H. Hersch, "Return of the Lost Spaceman: America's Astronauts in Popular Culture, 1959–2006," *The Journal of Popular Culture* 44 (2011): 73–92.
3. David Kaiser, "Cold War Requisitions, Scientific Manpower, and the Production of American Physicists after World War II," *Historical Studies in the Physical Sciences* 33, no. 1 (2002): 151–153.
4. Andrew Smith, *Moondust: In Search of the Men Who Fell to Earth* (New York: Fourth Estate, 2005); Thomas Mallon, "Moon Walker: How Neil Armstrong Brought the Space Program Down to Earth," *The New Yorker*, October 3, 2005.
5. David Kaiser, "The Postwar Suburbanization of American Physics," *American Quarterly* 56 (2004): 851–888.
6. Walter Cunningham, *The All-American Boys (Revised Edition)* (New York: ibooks, 2003), 285.
7. Compare, Kaiser, "The Postwar Suburbanization of American Physics."
8. Brian O'Leary, *The Making of an Ex-Astronaut* (Boston: Houghton Mifflin, 1970).
9. In general, though, military space service proved a wise gamble: pilots with only brief service as astronauts in the air force's MOL program did well in their later careers; one MOL pilot who left space service in 1969 enjoyed a prestigious military career as the head of President Ronald Reagan's Strategic Defense Initiative.
10. Alcestis R. Oberg, "Ten Years After: Most of the Members of NASA's 'Class of '78' Are Still Astronauts; Some Have Gone on to Other Jobs; And a Few Have Died," *Final Frontier*, October 1988, 35–37.
11. Franklin Musgrave, "World Space Week Lecture," Franklin Institute, Philadelphia, Pennsylvania, October 2, 2005; Henry W. Hartsfield, Jr., "Oral History Transcript #2 (Carol Butler, Interviewer)," NASA Johnson Space Center Oral History Project (Houston, Texas, June 15, 2001), 3.
12. Homer Hickam, "What Makes an Astronaut Crack," *Los Angeles Times*, February 9, 2007.
13. Clint Eastwood, *Firefox* (Warner Bros., 1982).
14. James L. Brooks, *Terms of Endearment* (Paramount Pictures, 1983).
15. Ralph Nader, *Unsafe at Any Speed: The Designed-In Dangers of the American Automobile* (New York: Grossman, 1965).
16. An ant colony built by high school students flew on a shuttle flight in 1983, but the ants within it died prior to liftoff. "Space Ants Died before Takeoff," *New York Times*, October 14, 1983, B2.
17. Carlos Baeza, "Deep Space Homer," *The Simpsons* (FOX, February 24, 1994).
18. Mimi Leder, *Deep Impact* (Paramount, 1998); Clint Eastwood, *Space Cowboys* (Warner Bros., 2000).
19. Donald H. Peterson, "Oral History Transcript (Jennifer Ross-Nazzal, Interviewer)," NASA Johnson Space Center Oral History Project (Houston, Texas, November 14, 2002), 77.

20. O'Leary, *The Making of an Ex-Astronaut*, 80.
21. Clint Eastwood, *Unforgiven* (Warner Bros., 1992).
22. Joseph D. Atkinson and Jay M. Shafritz, *The Real Stuff: A History of NASA's Astronaut Recruitment Program* (New York: Praeger, 1985), 193.
23. John Schwartz, "The New Astronaut: Plays Well with Robots," *New York Times*, August 14, 2005.

Selected Bibliography

The following sources were consulted extensively in the preparation of this work and will prove valuable to scholars seeking to explore this topic further.

Archives

George M. Low Papers. Archives and Special Collections, Rensselaer Polytechnic Institute, Troy, New York.

Johnson Space Center History Collection, University of Houston–Clear Lake, Texas.

NASA Historical Reference Collection, NASA Headquarters, Washington DC.

Records of the National Aeronautics and Space Adminisitration. National Archives and Records Center at College Park, Maryland.

Oral Histories

Allen, Joseph P. "Oral History Transcript (Jennifer Ross-Nazzal, Interviewer)." NASA Johnson Space Center Oral History Project. Houston, Texas, January 28, 2003.

———. "Oral History Transcript #2 (Jennifer Ross-Nazzal, Interviewer)." NASA Johnson Space Center Oral History Project. Washington DC, March 16, 2004.

———. "Oral History Transcript #3 (Jennifer Ross-Nazzal, Interviewer)." NASA Johnson Space Center Oral History Project. Washington DC, March 18, 2004.

Armstrong, Neil A. "Oral History Transcript (Dr. Stephen E. Ambrose and Dr. Douglas Brinkley, Interviewers)." NASA Johnson Space Center Oral History Project. Houston, Texas, September 19, 2001.

Bean, Alan L. "Oral History Transcript (Michelle Kelly, Interviewer)." NASA Johnson Space Center Oral History Project. Houston, Texas, June 23, 1998.

Bobko, Karol J. "Oral History Transcript (Summer Chick Bergen, Interviewer)." NASA Johnson Space Center Oral History Project. Houston, Texas, February 12, 2002.

Carpenter, M. Scott. "Oral History Transcript (Michelle Kelly, Interviewer)." NASA Johnson Space Center Oral History Project. Houston, Texas, March 30, 1998.

Cooper, L. Gordon, Jr. "Oral History Transcript (Roy Neal, Interviewer)." NASA Jet Propulsion Laboratory. Pasadena, California, May 21 1998.

Crippen, Robert L. "Oral History Transcript (Rebecca Wright, Interviewer)." NASA Johnson Space Center Oral History Project. Houston, Texas, May 26, 2006.

Fullerton, C. Gordon. "Oral History Transcript (Rebecca Wright, Interviewer)." NASA Johnson Space Center Oral History Project. NASA Dryden Flight Research Center, California, May 6, 2002.

Gibson, Edward G. "Oral History Transcript (Carol Butler, Interviewer)." NASA Johnson Space Center Oral History Project. Houston, Texas, December 1, 2000.

Gilruth, Robert. "Transcript #4 (Martin Collins, David DeVorkin, Interviewers)." Washington DC: Smithsonian Institution, National Air and Space Museum, October 2, 1986.

———. "Transcript #5 (David DeVorkin, John Mauer, Interviewers)." Washington DC: Smithsonian Institution, National Air and Space Museum, February 27, 1987.

———. "Transcript #6, Oral History Transcripts (David DeVorkin, John Mauer, Interviewers)." Washington DC: Smithsonian Institution, National Air and Space Museum, March 2, 1987.

Hartsfield, Henry W. Jr. "Oral History Transcript (Carol Butler, Interviewer)." NASA Johnson Space Center Oral History Project. Houston, Texas, June 12, 2001.

———. "Oral History Transcript #2 (Carol Butler, Interviewer)." NASA Johnson Space Center Oral History Project. Houston, Texas, June 15, 2001.

Kerwin, Joseph P. "Oral History Transcript (Kevin M. Rusnak, Interviewer)." NASA Johnson Space Center Oral History Project. Houston, Texas, May 12, 2000.

Mattingly, Thomas Kenneth II. "Oral History Transcript #2 (Kevin M. Rusnak, Interviewer)." NASA Johnson Space Center Oral History Project. Houston, Texas, April 22, 2002.

Peterson, Donald H. "Oral History Transcript (Jennifer Ross-Nazzal, Interviewer)." NASA Johnson Space Center Oral History Project. Houston, Texas, November 14, 2002.

Schirra, Walter M. Jr. "History Transcript (Roy Neal, Interviewer)." NASA Johnson Space Center Oral History Project. San Diego, California, December 1, 1998.

Thompson Milton O. and Curtis Peebles. *Flying Without Wings: NASA Lifting Bodies and the Birth of the Space Shuttle.* Washington DC, Smithsonian Institution Press, 1999.

Truly, Richard H. "Oral History Transcript (Rebecca Wright, Interviewer)." NASA Oral History Project. Golden, Colorado, June 16, 2003.

Newspaper and Magazine Articles

"Astronaut Resigns to Pursue Science." *New York Times*, August 6, 1969, 18.

"Astronauts Push for a 7th Flight." *New York Times*, May 22, 1963, 1.

Bledsoe, Jerry. "Down from Glory." *Esquire*, January 1973, 83–86.

Borenstein, Seth. "Astronauts Recall View before Earth Day: Space Travelers Recall What It's Like to See Their Home Planet from above Ahead of Earth Day." *Associated Press*, April 21, 2007.

Chaikin, Andrew. "George Abbey: NASA's Most Controversial Figure." *Space.com*, February 26, 2001.

Finney, John W. "7 Named as Pilots for Space Flights Scheduled in 1961." *New York Times*, April 10, 1959, 1.

Goldstein, Richard. "Walter M. Schirra Jr., Astronaut, Dies at 84." *New York Times*, May 4, 2007, B7.

Grimes, William. "Joan Winston, 'Trek' Superfan, Dies at 77." *New York Times*, September 21, 2008, 34.

Henig, Robin Marantz. "Understanding the Anxious Mind." *New York Times Magazine*, October 4, 2009, 30–64.

Hickam, Homer. "What Makes an Astronaut Crack," *Los Angeles Times*, February 9, 2007, A25.
Hill, Arthur. "Scientist-Astronauts Facing Uncertain Future." *Houston Chronicle*, May 23, 1971, Section 4, 7.
Hill, Gladwin. "Test Pilots Get Some Good News." *New York Times*, October 9, 1959, 12.
Howard, William E. "USIA Editors Scooped on Flight Stories." *The Birmingham News*, May 29, 1963, 16.
Ivins, Molly. "Ed Who?" *New York Times*, June 30, 1974, 12–16.
Lapp, Ralph E. "Send Computers, Not Men, into Deep Space." *New York Times*, February 2, 1969, SM32.
Leary, Warren E. "NASA Opens Inquiry into Drunken-Flying Reports." *New York Times*, July 28, 2007, 11.
Loudon, James A. "Why We Really Want a Space Shuttle." *New York Times*, March 28, 1972, 42.
Lyons, Richard D. "Each Astronaut Is an Only Child: Fact Cited as Evidence of Theory of Achievement." *New York Times*, December 24, 1968, 7.
Mallon, Thomas. "Moon Walker: How Neil Armstrong Brought the Space Program Down to Earth." *The New Yorker*, October 3, 2005, 94–98.
———. "Satellite of Love." *New York Times*, April 9, 2006, 7.
——— "The Moon Landing." *New York Times*, July 18, 2009, A20.
Muson, Howard. "Comedown from the Moon: What Has Happened to the Astronauts." *New York Times Magazine*, December 3, 1972, 37–140.
O'Leary, Brian. "Topics: Science or Stunts on the Moon?" *New York Times*, April 25, 1970, 18.
———. "Do We Really Want a Space Shuttle?" *New York Times*, February 16, 1972, 39.
———. "Space Shuttle: Billions for a Boondoggle?" *New York Times*, May 16, 1972, 42.
O'Toole, Thomas. "NASA Gives Police Power to Spaceship Captains." *Washington Post*, March 18, 1980, A8.
Oberg, Alcestis R. "Ten Years After: Most of the Members of NASA's 'Class of '78' Are Still Astronauts. Some Have Gone on to Other Jobs. And a Few Have Died." *Final Frontier*, October 1988, 35–37.
Reinhold, Robert "Shuttle Mission Puts Focus on Research Crewmen." *New York Times*, November 13, 1982, 10.
Reston, James. "Washington: The Sky's No Longer the Limit." *New York Times*, April 12, 1959, E8.
Richard D. Lyons. "Each Astronaut Is an Only Child: Fact Cited as Evidence of Theory of Achievement." *New York Times*, December 24, 1968, 7.
——— "Men in Space." *New York Times*, April 11, 1959, 20.
Schwartz, John. "The New Astronaut: Plays Well with Robots." *New York Times*, August 14, 2005, C4.
———. "Astronauts Have Flown While Drunk, NASA Finds." *New York Times*, July 27, 2007, A16.
——— "Science: Commuting in Space." *Time*. December 15, 1975, 50–53.
Sherrod, Robert. "The Selling of the Astronauts." *Columbia Journalism Review*, May/June 1973, 16–25.
———"Shuttle Crew Selection Change Worries Scientific Community." *Aviation Week & Space Technology*, August 7, 1978, 23.
———"Space Ants Died before Takeoff." *New York Times*, October 14, 1983, B2.
Wilford, John Noble. "Robert Gilruth, 86, Dies: Was Crucial Player at NASA." *New York Times*, August 18, 2000, C19.

Witkin, Richard. "Scientist Expected to Be Picked for Moon Trip: Space Agency Reported Set to Name a Geologist Move Is Viewed as Attempt to Answer Criticism." *New York Times*, December 12, 1969, 33.

Secondary Sources

Abbott, Andrew Delano. *The System of Professions: An Essay on the Division of Expert Labor.* Chicago: University of Chicago Press, 1988.

Aldrin, Edwin E., and Wayne Warga. *Return to Earth.* New York: Random House, 1973.

Anker, Peder. "The Ecological Colonization of Space." *Environmental History* 10 (2005): 239–268.

Atkinson, Joseph D. and Jay M. Shafritz. *The Real Stuff: A History of NASA's Astronaut Recruitment Program.* New York: Praeger, 1985.

Ben-David, Joseph. *The Scientist's Role in Society: A Comparative Study.* Englewood Cliffs: Prentice-Hall, 1971.

Beyer, D. H. and S. B. Sells. "Selection and Training of Personnel for Space Flight." *Journal of Aviation Medicine* 28 (1957): 1–6.

Boushey, Homer A. "Blueprints for Space." In *Man in Space: The United States Air Force Program for Developing the Spacecraft Crew*, edited by Kenneth Franklin Gantz, 238–253. New York: Duell, 1959.

Bowles, Mark D. *Science in Flux: NASA's Nuclear Program at Plum Brook Station, 1955–2005.* Washington DC: NASA, 2006.

Bradbury, Ray. *The Illustrated Man.* Garden City: Doubleday & Company, 1951.

Braverman, Harry. *Labor and Monopoly Capital: The Degradation of Work in the Twentieth Century.* New York: Monthly Review Press, 1975.

Brooks, Courtney G., James M. Grimwood, and Loyd S. Swenson. *Chariots for Apollo: A History of Manned Lunar Spacecraft.* Washington: NASA, 1979.

Buckbee, Ed and Wally Schirra. *The Real Space Cowboys.* Burlington, Ontario: Apogee Books, 2005.

Burgess, Colin. *Selecting the Mercury Seven: The Search for America's First Astronauts.* Chichester: Springer-Praxis, 2011.

Caidin, Martin. *Marooned.* New York: Dutton, 1964.

Calvert, Monte A. *The Mechanical Engineer in America, 1830–1910: Professional Cultures in Conflict.* Baltimore: Johns Hopkins Press, 1967.

Campbell-Kelly, Martin and William Aspray. *Computer: A History of the Information Machine.* Boulder: Westview Press, 2004.

Carpenter, M. Scott, et al. *We Seven.* New York: Simon and Schuster, 1962.

——— and Kris Stoever. *For Spacious Skies: The Uncommon Journey of a Mercury Astronaut.* Orlando: Harcourt, 2002.

Catchpole, John. *Project Mercury: NASA's First Manned Space Programme*, Springer-Praxis Books in Astronomy and Space Sciences. Chichester: Praxis, 2001.

Cernan, Eugene and Don Davis. *The Last Man on the Moon: Astronaut Eugene Cernan and America's Race in Space.* New York: St. Martin's Press, 1999.

Chaiken, Andrew. A Man on the Moon: The Voyages of the Apollo Astronauts. New York: Viking, 1994.

Cohn, Carol. "Wars, Wimps, and Women: Talking Gender and Thinking War." In *Gendering War Talk*, edited by Miriam Cooke and Angela Woollacott, 227–246. Princeton: Princeton University Press, 1993.

Collins, Michael. *Carrying the Fire: An Astronaut's Journeys.* New York: Farrar, 1974.

Compton, W. David and Charles D. Benson. *Living and Working in Space: A History of Skylab*. Washington DC: NASA, 1983.

Conway, Erik M. and Edward Potts Cheyney Memorial Fund. *High-Speed Dreams: NASA and the Technopolitics of Supersonic Transportation, 1945–1999*, New Series in NASA History. Baltimore: Johns Hopkins University Press, 2005.

Cox, Stephen. *Dreaming of Jeannie: TV's Prime Time in a Bottle*. New York: St. Martin's Griffin, 2000.

Cunningham, Walter. *The All-American Boys (Revised Edition)*. New York: ibooks, 2003.

——— and Mickey Herskowitz. *The All-American Boys*. New York: Macmillan, 1977.

De Groot, Gerard J. *Dark Side of the Moon: The Magnificent Madness of the American Lunar Quest*. New York: New York University Press, 2006.

Edwards, Richard. *Contested Terrain: The Transformation of the Workplace in the Twentieth Century*. New York: Basic Books, 1979.

Ezell, Edward Clinton and Linda Neuman Ezell. *The Partnership: A History of the Apollo-Soyuz Test Project*. Washington DC: NASA, 1978.

Farquhar, Robert. *Fifty Years on the Space Frontier: Halo Orbits, Comets, Asteroids, and More*. Parker: Outskirts Press, 2011.

Foster, Amy Elizabeth. Integrating Women into the Astronaut Corps: Politics and Logistics, 1972–2004. Baltimore: Johns Hopkins University Press, 2011.

French, Francis and Colin Burgess. *Into That Silent Sea: Trailblazers of the Space Era, 1961–1965*, Outward Odyssey. Lincoln: University of Nebraska Press, 2007.

———. *In the Shadow of the Moon: A Challenging Journey to Tranquility, 1965–1969*. Lincoln: University of Nebraska Press, 2007.

Fries, Sylvia Doughty. *NASA Engineers and the Age of Apollo*. Washington DC: NASA, 1992.

Fuller, R. Buckminster. *Operating Manual for Spaceship Earth*. Carbondale: Southern Illinois University Press, 1969.

Gantz, Kenneth Franklin, ed. *Man in Space: The United States Air Force Program for Developing the Spacecraft Crew*. New York: Duell, 1959.

Gerovitch, Slava. " 'New Soviet Man' inside the Machine: Human Engineering, Spacecraft Design, and the Construction of Communism." *Osiris* 22 (2007): 135–157.

Godwin, Robert, ed. *Dyna-Soar: Hypersonic Strategic Weapons System*. Burlington, Ontario: Apogee Books, 2003.

Grissom, Betty and Henry Still. *Starfall*. New York: Crowell, 1974.

Grosser, George H., Henry Wechsler and Milton Greenblatt, ed. *The Threat of Impending Disaster, Contributions to the Psychology of Stress*. Cambridge, Massachusetts: MIT Press, 1964.

Hansen, James R. *First Man: The Life of Neil A. Armstrong*. New York: Simon & Schuster, 2005.

Haraway, Donna Jeanne. *Primate Visions: Gender, Race, and Nature in the World of Modern Science*. New York: Routledge, 1989.

Heppenheimer, T. A. *Development of the Shuttle, 1972–1981*. Washington: Smithsonian Institution Press, 2002.

Hersch, Matthew H. "Space Madness: The Dreaded Disease that Never Was." *Endeavour* 36 (2012): 32–40.

———. "Return of the Lost Spaceman: America's Astronauts in Popular Culture, 1959–2006." *The Journal of Popular Culture* 44 (2011): 73–92.

———. "Checklist: The Secret Life of Apollo's 'Fourth Crewmember'." In *Space Travel & Culture: From Apollo to Space Tourism*, edited by David Bell and Martin Parker, 6–24. Oxford, UK: Blackwell, 2009.

Horwitch, Mel. *Clipped Wings: The American SST Conflict.* Cambridge, Massachusetts: MIT Press, 1982.

Irwin, James B. *More Than Earthlings: An Astronaut's Thoughts for Christ-Centered Living.* Nashville: Broadman Press, 1983.

Jasanoff, Sheila. "Image and Imagination." In *Changing the Atmosphere: Expert Knowledge and Environmental Governance*, edited by Clark A. Miller and Paul N. Edwards, 309–337. Cambridge, Massachusetts: MIT Press, 2001.

Johnson, Stephen B. *The Secret of Apollo: Systems Management in American and European Space Programs.* Baltimore: Johns Hopkins University Press, 2002.

Kaiser, David. "Cold War Requisitions, Scientific Manpower, and the Production of American Physicists after World War II." *Historical Studies in the Physical Sciences* 33, no. 1 (2002): 131–159.

———. "The Postwar Suburbanization of American Physics." *American Quarterly* 56, (2004): 851–888.

Keniston, Kenneth. *Youth and Dissent: the Rise of a New Opposition.* New York: Harcourt Brace Jovanovich, 1971.

Kevles, Bettyann. *Almost Heaven: The Story of Women in Space.* New York: Basic Books, 2003.

Kilgore, De Witt Douglas. *Astrofuturism: Science, Race, and Visions of Utopia in Space.* Philadelphia: University of Pennsylvania Press, 2003.

Kohler, Robert. *Lords of the Fly: Drosophila Genetics and the Experimental Life.* Chicago: University of Chicago Press, 1994.

Kraft, Christopher C. *Flight: My Life in Mission Control.* New York: Dutton, 2001.

Kunda, Gideon. *Engineering Culture: Control and Commitment in a High-Tech Corporation.* Philadelphia: Temple University Press, 1992.

Lambright, W. Henry. *Powering Apollo: James E. Webb of NASA.* Baltimore: Johns Hopkins University Press, 1995.

Larson, Magali Sarfatti. *The Rise of Professionalism: A Sociological Analysis.* Berkeley: University of California Press, 1977.

Latour, Bruno and Steve Woolgar. *Laboratory Life: The Construction of Scientific Facts.* Princeton: Princeton University Press, 1986.

Launius, Roger D. *NASA: A History of the U.S. Civil Space Program.* Malabar, Florida: Krieger Pub. Co., 1994.

———. "NASA History and the Challenge of Keeping the Contemporary Past." *The Public Historian* 21, no. 3 (1999): 63–81.

Levine, Arnold S. *Managing NASA in the Apollo Era.* Washington DC: NASA, 1982.

Lewallen, Constance M. and Steve Seid. *Ant Farm 1968–1978.* Berkeley: University of California Press, 2004.

Licht, Walter. *Industrializing America: The Nineteenth Century*, American Moment. Baltimore: Johns Hopkins University Press, 1995.

Link, Mae Mills. *Space Medicine in Project Mercury.* Washington DC: NASA, 1965.

Logsdon, John M., ed. *Exploring the Unknown: Selected Documents in the History of the U.S. Civil Space Program, Volume I.* Washington DC: NASA, 1995.

———. *Exploring the Unknown: Selected Documents in the History of the U.S. Civil Space Program, Volume VII.* Washington DC: NASA, 2008.

Mack, Pamela Etter. *Viewing the Earth: The Social Construction of the Landsat Satellite System.* Cambridge, Massachusetts: MIT Press, 1990.

Mackowski, Maura Phillips. *Testing the Limits: Aviation Medicine and the Origins of Manned Space Flight.* College Station, Texas: Texas A&M University Press, 2006.
Maher, Neil. "Neil Maher on Shooting the Moon." Environmental History 9 (2004): 526–531.
Mailer, Norman. *Of a Fire on the Moon.* Boston: Little, Brown, 1970.
Manned Operations Branch, Flight Crew Support Division. "Apollo 11 Technical Crew Debriefing (U), Vol. 1." Houston: NASA, 1969.
———. "Apollo 11 Technical Crew Debriefing (U), Vol. 2." Houston: NASA, 1969.
McCurdy, Howard E. *Inside NASA: High Technology and Organizational Change in the U.S. Space Program.* Baltimore: Johns Hopkins University Press, 1993.
———. *Space and the American Imagination*, Smithsonian History of Aviation Series. Washington DC: Smithsonian Institution Press, 1997.
McDougall, Walter A. *The Heavens and the Earth: A Political History of the Space Age.* Baltimore: Johns Hopkins University Press, 1997.
Mclaughlin, E. J. "Family Structure of Astronauts." Houston: Space Medicine, NASA Manned Space Flight Center, 1969.
Mindell, David A. "Human and Machine in the History of Spaceflight." In *Critical Issues in the History of Spaceflight*, edited by Steven J. Dick and Roger D. Launius, 141–168. Washington DC: NASA, 2006.
———. *Digital Apollo: Human and Machine in Spaceflight.* Cambridge, Massachusetts: MIT Press, 2008.
Morse, Mary Louise and Jean Kernahan Bays. *Apollo Spacecraft: A Chronology, Volume II.* Washington DC: NASA, 1973.
Mukerji, Chandra. *A Fragile Power: Scientists and the State.* Princeton: Princeton University Press, 1989.
Mullane, R. Mike. *Riding Rockets: The Outrageous Tales of a Space Shuttle Astronaut.* New York: Scribner, 2006.
Nader, Ralph. *Unsafe at Any Speed, the Designed-in Dangers of the American Automobile.* New York: Grossman, 1965.
National Aeronautics and Space Administration. *Apollo 11 Technical Air-to-Ground Voice Transcription (GOSS NET 1).* Houston: Manned Spacecraft Center, 1969.
———. *Astronaut Fact Book.* Houston: NASA Johnson Space Center, 2005.
Needell, Allan A. *Science, Cold War and the American State: Lloyd V. Berkner and the Balance of Professional Ideals.* Amsterdam: Harwood Academic Publishers, 2000.
Nelson, Bruce. *Divided We Stand: American Workers and the Struggle for Black Equality*, Politics and Society in Twentieth-Century America. Princeton: Princeton University Press, 2001.
Neufeld, Michael J. *Von Braun: Dreamer of Space, Engineer of War.* New York: A.A. Knopf, 2007.
Nicogossian, Arnauld E., Carolyn Leach Huntoon, and Sam L. Pool. *Space Physiology and Medicine.* Philadelphia: Lea & Febiger, 1994.
Noble, David F. *Forces of Production: A Social History of Industrial Automation.* New York: Knopf, 1984.
Nye, David E. *American Technological Sublime.* Cambridge, Massachusetts: MIT Press, 1994.
O'Leary, Brian. *The Making of an Ex-Astronaut.* Boston: Houghton Mifflin, 1970.
Papazian, Ed. *Medium Rare: The Evolution, Workings, and Impact of Commercial Television.* New York: Media Dynamics, 1991.

Philmus, Lois C. *A Funny Thing Happened on the Way to the Moon*. New York: Spartan Books, 1966.

Portola Institute. *The Last Whole Earth Catalog: Access to Tools*. Random House, New York, 1971.

Portree, David S. F. *Humans to Mars: Fifty Years of Mission Planning, 1950–2000*, Monographs in Aerospace History. Washington DC: NASA, 2001.

Reich, Charles A. *The Greening of America*. New York: Random House, 1970.

Rozwadowski, Helen M. "Small World: Forging a Scientific Maritime Culture for Oceanography." *Isis* 87, (1996): 409–429.

Ruff, George E. and Edwin Z. Levy. "Psychiatric Evaluation of Candidates for Space Flight." *American Journal of Psychiatry* 116, (1959): 385–391.

Ryan, Craig. *The Pre-Astronauts: Manned Ballooning on the Threshold of Space*. Annapolis: Naval Institute Press, 1995.

Santy, Patricia A. *Choosing the Right Stuff: The Psychological Selection of Astronauts and Cosmonauts*. Westport: Praeger, 1994.

Schefter, James L. *The Race: The Uncensored Story of How America Beat Russia to the Moon*. New York: Doubleday, 1999.

Schirra, Wally and Richard N. Billings. *Schirra's Space*. Boston: Quinlan Press, 1988.

Schivelbusch, Wolfgang. *The Railway Journey: Trains and Travel in the 19th Century*. New York: Urizen Books, 1979.

Scott, Joan Wallach. "Gender: A Useful Category of Historical Analysis." In *Gender and the Politics of History*. New York: Columbia University Press, 1999.

Shayler, David J. *Skylab: America's Space Station*, Springer-Praxis Books in Astronomy and Space Sciences. Chichester, UK: Praxis, 2001.

———. *Apollo: The Lost and Forgotten Missions*. Chichester, UK: Springer, 2002.

———. *NASA's Scientist-Astronauts*. New York: Springer, 2007.

Siddiqi, Asif A. *Challenge to Apollo: The Soviet Union and the Space Race, 1945–1974*. Washington DC: NASA, 2000.

———. *The Rockets' Red Glare: Spaceflight and the Russian Imagination, 1857–1957*. Cambridge, UK: Cambridge University Press, 2010.

Siegel, Dick. "Rip Van Spaceman: First American Spaceman Returns after 47 Years!" *Weekly World News*, July 25, 2005.

Slayton, Donald K. and Michael Cassutt. *Deke! U.S. Manned Space: From Mercury to the Shuttle*. New York: St. Martin's Press, 1994.

Smith, Andrew. *Moondust: In Search of the Men Who Fell to Earth*. New York: Fourth Estate, 2005.

Smith, Merritt Roe. *Harpers Ferry Armory and the New Technology: The Challenge of Change*. Ithaca, New York: Cornell University Press, 1977.

Smith, Michael L. *Pacific Visions: California Scientists and the Environment, 1850–1915*. New Haven, Connecticut: Yale University Press, 1987.

Sobchack, Vivian. "The Virginity of Astronauts: Sex and the Science Fiction Film." In *Alien Zone: Cultural Theory and Contemporary Science Fiction Cinema*, edited by Annette Kuhn, 103–115. New York: Verso, 1990.

Society of Experimental Test Pilots. *History of the First 20 Years*. Covina: Taylor Pub. Co., 1978.

Stafford, Thomas P. and Michael Cassutt. *We Have Capture: Tom Stafford and the Space Race*. Washington DC: Smithsonian Institution Press, 2002.

Steele, Allen M. *Orbital Decay*. New York: Ace Books, 1989.

Swanson, Glen E., ed. *Before This Decade Is Out: Personal Reflections on the Apollo Program*. Washington DC: NASA, 1999.

Swenson, Loyd S., James M. Grimwood, and Charles C. Alexander. *This New Ocean, a History of Project Mercury*. Washington DC: NASA, 1966.
Thompson, Neal. *Light This Candle: The Life and Times of Alan Shepard, America's First Spaceman*. New York: Crown Publishers, 2004.
Tomayko, J. E. *Computers in Space: Journeys with NASA*. Indianapolis: Alpha Books, 1994.
Traweek, Sharon. *Beamtimes and Lifetimes: The World of High Energy Physicists*. Cambridge, Massachusetts: Harvard University Press, 1988.
Turner, Fred. *From Counterculture to Cyberculture: Stewart Brand, the Whole Earth Network, and the Rise of Digital Utopianism*. Chicago: University of Chicago Press, 2006.
Van Riper, Frank. *Glenn, the Astronaut Who Would Be President*. New York: Empire Books, 1983.
Vaughan, Diane. *The Challenger Launch Decision: Risky Technology, Culture, and Deviance at NASA*. Chicago: University of Chicago Press, 1996.
Verne, Jules. *From the Earth to the Moon: and, a Trip around It*. Philadelphia: Lippincott, 1950.
Von Braun, Wernher and Cornelius Ryan. *Conquest of the Moon*. New York: Viking Press, 1953.
Weitekamp, Margaret A. *Right Stuff, Wrong Sex: America's First Women in Space Program*. Baltimore: Johns Hopkins University Press, 2004.
Wells, Robert. *What Does an Astronaut Do?* New York: Dodd, 1961.
Wendt, Guenter and Russell Still. *The Unbroken Chain*. Burlington, Ontario: Apogee Books, 2001.
Wilson, Charles L., ed. *Project Mercury Candidate Evaluation Program (Technical Report 59–505)*. Dayton: Wright Air Development Center, 1959.
Wilson, Sloan. *The Man in the Gray Flannel Suit*. New York: Simon and Schuster, 1955.
Winston, Joan. *The Making of the Trek Conventions: Or, How to Throw a Party for 12,000 of Your Most Intimate Friends*. New York: Doubleday, 1977.
Wolfe, Peter. *In the Zone: The Twilight World of Rod Serling*. Bowling Green, Ohio: Bowling Green State University Popular Press, 1997.
Wolfe, Tom. *The Right Stuff*. New York: Farrar, Straus, and Giroux, 1979.
Worden, Al. *Falling to Earth: An Apollo Astronaut's Journey*. Washington DC: Smithsonian Books, 2011.
Zussman, Robert. *Mechanics of the Middle Class: Work and Politics among American Engineers*. Berkeley: University of California Press, 1985.

Motion Pictures and Television

Allen, Woody. "Bananas." United Artists, 1971.
Baeza,Carlos. "Deep Space Homer." *The Simpsons*. FOX, February 24, 1994.
Blatty,William Peter. "The Ninth Configuration." Warner Bros., 1980.
Brooks, James L. "Terms of Endearment." Paramount Pictures, 1983.
Eastwood, Clint. "Firefox." Warner Bros., 1982.
———. "Unforgiven." Warner Bros., 1992.
———. "Space Cowboys." Warner Bros., 2000.
Friedkin, William. "The Exorcist." Hoya Productions, 1973.
Gilbert, Lewis. "You Only Live Twice." United Artists, 1967.
———. "Moonraker." United Artists, 1979.

Haskin, Byron. "Conquest of Space." Paramount Pictures Corp., 1955.
———. "Robinson Crusoe on Mars." Paramount Pictures Corp., 1964.
Kaufman, Philip. "The Right Stuff." The Ladd Company, 1983.
Leder, Mimi. "Deep Impact." Paramount, 1998.
Medford, Don. "Death Ship." *Twilight Zone.* CBS, February 7, 1963.
Neumann, Kurt. "Rocketship X-M." Lippert Pictures, 1950.
Sington, David. "In the Shadow of the Moon." Discovery Films, 2007.
Sturges, John. "Marooned." Columbia Pictures Corp., 1969.
Tarkovsky, Andrei. "Solyaris." Mosfilm, 1972.
Trumbull, Douglas. "Silent Running." Universal Pictures, 1972.

Dissertations and Theses

Sato, Yasushi. "Local Engineering in the Early American and Japanese Space Programs: Human Qualities in Grand System Building." PhD diss., University of Pennsylvania, 2005.
Starr, Kristen. "NASA's Hidden Power: NACA/NASA Public Relations and the Cold War, 1945–1967." PhD diss., Auburn University, 2008.

Public Presentations

Aldrin, Buzz. "Remarks Accompanying Screening of in The Shadow of the Moon." Heritage Foundation, Washington DC:, September 7, 2007.
Borman, Frank, Jim Lovell, and Bill Anders. "John H. Glenn Lecture: An Evening with the Apollo 8 Astronauts." Smithsonian Institution, National Air and Space Museum, Washington DC, November 13, 2008.
Launius, Roger D. "Heroes in a Vacuum: The Apollo Astronaut as Cultural Icon." Paper presented at the 43rd AIAA Aerospace Sciences Meeting and Exhibit, Reno, Nevada, January 10–13, 2005.
Levasseur, Jennifer. " 'Here's the Earth Coming up': Analysis of the Apollo 8 'Earthrise' Photograph." Paper presented at the Annual Meeting of the Society for the History of Technology, Lisbon, Portugal, October 12, 2008.
Musgrave, Franklin. "World Space Week Lecture." Franklin Institute, Philadelphia, Pennsylvania, October 2, 2005.
Walz, Carl. "Astronaut Adventures." Paper presented at the Smithsonian Institution Folklife Festival, Washington DC, June 26, 2008.

Index

Page numbers in *italics* denote figures.